ANCIENT COINS ANCIENT SKIES

Ancient Coins
Ancient Skies

·

PAPERS & ARTICLES

GEORGE LATURA

To Boopsie,
– "Beautiful *and* smart."

CONTENTS

INTRODUCTION

These articles and papers might serve as a preliminary roadmap to a renewed appreciation of the ancient world. Just as the first Christian prayer invoked "Our Father who art in heaven," so did the earlier cosmology look to the skies, with the Wanderers embodying Olympian gods – Jupiter, Sol, Venus, Luna, etc. – whose divine bounty flowed down from the heavens as the Cornucopia, along the Milky Way.

After its travails here on earth, a just soul would rejoin the gods in the heavens, just as Caesar's soul sailed skyward as a comet during his funeral games in 44 BC, setting the stage for the Roman Empire. Up along the ladder of the Planets climb those who fall in battle – as shown on the standards at the fore of Roman armies – to be greeted by Divus Julius and later Divus Augustus at the heavenly gates.

Heaven itself was the Milky Way, where the gods and heroes and sages resided – according to Heraclides of Pontum, Cicero, Ovid, Manilius, Numenius, Macrobius, Martianus Capella, etc. At the intersections of the Milky Way and the path of the Planets (Zodiac) stood the celestial portals, as revealed by Macrobius in his *Commentary on Cicero's Dream of Scipio*, which can be traced back to Plato's Myth of Er at the end of *Republic*.

The celestial intersection where stood the gates of heaven would appear on Roman coins over centuries, from the Republic to the end of the Empire.

Constantine's True Vision

FROM PLATO'S *CHI* (X) TO THE CHRISTIAN *CHI RHO*

COIN NEWS numismatic monthly (UK), February 2013

NUMISMATIC LITERARY GUILD AWARD – Best Article 2013 (International Numismatic Magazines)

"THIS IS NOT THE WAY THINGS were supposed to go," thought Constantine. Too weak to raise a finger, the ruler of the Roman Empire watched helplessly as the Christian bishop barged past his imperial guard, clutching vials of magic water and oil.

"In the name of the Father, and the Son, and – " One, two, three! The Emperor of Rome – Pontifex Maximus of the Roman state religion – was made into a Christian just as he gave up his soul.

Later bishops would sweep Constantine's atrocities and murders under the ecclesiastical rug, claiming that the emperor had been a fervent Christian all along. Brandishing this fabrication, they pounced on the 'pagan' culture that had built and sustained the Hellenistic world, and they destroyed its temples, burned its books, and declared the performance of ancient rites punishable by death. With this cultural lobotomy, much of Europe plunged into a thousand years of ignorance, illiteracy, and intolerance that we now call the Dark Ages.

But that's not what Constantine had envisioned. Constantine had dreamed of a marriage, a union of the thousand-year-old civilization that had nurtured the golden age of Greece and Rome and the radical precepts of a New Age cult – an offshoot of Judaism – that was sweeping the provinces. Erased from text and memory by Christian censors over the centuries, the unifying vision of Constantine still shines through on coins of that tumultuous era.

The Stairway to Heaven

Just as they had adorned coins of Roman emperors over hundreds of years, pagan planetary gods appeared on early coins struck by Constantine: Mars marches with him on his military expeditions, the king of the gods Jupiter presents the celestial sphere that stands for command over the cosmos, and Sol Invictus offers the same cosmic orb to the emperor on numerous coins. To the Roman army, Jupiter, Mars and the Sun were of course the most powerful, visible gods in the heavens.

According to savants of the time, the Wanderers in the sky travel along the ecliptic and trace out the Zodiac, the constellations that predict how a man's life will proceed and eventually end. And when they are fortuitously aligned, the Planets illustrate step-by-step the visible stairway to heaven.

The planetary ladder to the heavens was carried at the forefront of legions on Roman standards, promising a heavenly afterlife to any soldier who might die that day in the field, a pledge that Constantine guaranteed, in his role of emperor and highest priest (Figure 1).

4

Figure 1. Coin of Constantine with standards of the Roman army showing the planetary ladder to heaven.

Most popular in the Roman army, the cult of Mithras preached the ascent to the heavens through the Planets, as the Greek philosopher Celsus described:

> In that system there is an orbit for the fixed stars, another for the planets and a diagram for the passage of the soul through the latter. They picture this as a ladder with seven gates, and at the very top an eighth gate... – Celsus, On the True Doctrine

The celestial ladder can be found on every Roman coin that sports a legionary standard with disks or circles that represent the Planets and their orbits ('Celestial Symbols on Roman Standards,' The Celator, June 2011).

The Gates of Heaven

Where does the heavenly stairway of the Planets take us? To the crossroads in the sky, to the intersections of the Milky Way and the course of the Planets, where the gates of heaven stood according to Plato's disciples (who quote his Republic and Timaeus) and as seen on Roman coins for centuries ('Plato's X on Roman Coins,' COIN NEWS, January 2012).

The X in the sky was not a late addition to Roman cosmology, for we can already find it on coins of the Republic. A denarius struck by Manlius shows the celestial X sitting by the Sun in his celestial quadriga, who is flanked by the crescent Moon – these being the two most prominent Wanderers in the skies (Figure 2).

5

Figure 2. Republic denarius of Manlius (c. 107 BC), showing on the reverse the Sun in his quadriga, the crescent Moon, and on the left, the celestial intersection, Plato's X in the sky.

Who controlled the Gates of Heaven? The Pontifex Maximus did – the highest priest of the state cult of Rome – in the position held by Julius Caesar, by Augustus, and by every Roman emperor thereafter. In the grasp of the gods sits the celestial sphere, whether offered to the emperor by Jupiter or by Sol Invictus, and on that heavenly orb we see the intersecting lines that indicate the heavenly portals. That is the divine right by which Constantine ruled, the gods having granted him control of the celestial gates that sit at the crossroads in the sky (Figure 3).

Figure 3. Coin of Constantine with Sol Invictus granting the emperor command over the gates of heaven at the celestial X.

As Pontifex Maximus, the ruler of Rome guaranteed to the citizens of the Empire that he was in intimate contact with the cosmic powers through the celestial portals, and thus could

6

most reliably wield the rudder that steered the world to its shining future.

The Planets and the Celestial Crossroads

Constantine's genius led him to combine two celestial symbols into one. The climb up the ladder of the Planets leads to the gates of heaven – the crossroads in the sky that had appeared on Roman coins for centuries. This combined symbolism was given a Christian twist when, by the addition of a vertical crossbar, Plato's celestial Chi (X) would be transformed into the Chi Rho, cementing the union of the pagan and the Christian worldviews (Figure 4).

One rarely finds Christian symbols on Constantine's own coins, but the Chi Rho banner atop the planetary ladder does appear toward the end of his reign on coins of junior emperors – his sons Constantine Jr., Constantius II, and his nephew Delmatius.

Figure 4. Coin of Delmatius, junior emperor under Constantine, with the planetary ladder leading up to the heavenly gates now marked by the Chi Rho (c. 336 AD). Once Constantine passed away, his Christian sons had Delmatius and other cousins killed, following their father's practice of eliminating any possible competition.

Having made themselves co-emperors, the three brothers embraced the combined symbolism of the planetary stairs that led to the gates of heaven now indicated by the Christian Chi Rho. In this merged cosmic view, the planetary gods of the pagan world still had a role in the scheme of things, serving as stepping-stones to the celestial portals (Figure 5).

7

Figure 5. Coin of Constantius II, with the emperor holding standards with planetary ladder and Chi Rho at the top.

Constantine's wedding of pagan and Christian symbolism was not without precedent. Already around 165 AD, the Christian apologist Justin Martyr had tried that tactic.

Re-branding the Old, Erasing the Past

In an open letter to the emperor Antoninus Pius, Justin equated the Christian Son of God with Plato's Cosmic Soul, which had the shape of an X in the sky.

> And the scientific discussion of the Son of God in his *Timaeus* – when he says: "He arranged him as an X in the cosmos" – Plato took from Moses, and spoke in similar terms. – Justin Martyr, *Apology on Behalf of Christians*

Here was the first step in the re-branding of Plato's ancient symbol, the celestial X that had appeared on many Roman coins: Justin claimed that Plato was discussing the Son of God 350 years before Jesus was born, and that Plato had somehow purloined the shape of the cross from Moses. The Christian upstarts were trying to grab control of the gates of heaven from the very hands of the Roman Emperor. No wonder that Justin soon earned his nickname of 'martyr.'

Constantine wasn't so picky. Like earlier bootstrap emperors such as Augustus and Vespasian, he needed to establish a heavenly mandate for the dynasty of his ambitions, and he would take his celestial endorsements from whatever corner they might come from.

In fact, the first divine vision that blessed Constantine's ambitions came from a pagan god.

> On his way either to or back from Massilia, Constantine received news of the final collapse of the barbarian uprising on the Rhine. The news was conveyed at the precise point of the journey at which there was a road leading to a sanctuary of Apollo described by the panegyrist as 'the most beautiful temple in the whole world.' It was there, according to the panegyrist in the climactic part of his speech, that the god himself appeared to the emperor, accompanied by Victory... This first recorded and purely pagan religious experience of Constantine has been seen by some modern scholars as 'the only authentic vision of Constantine, the legend of the vision of 312 being nothing but a Christian distortion.' – Samuel N.C. Lieu and Dominic Monserrat, *From Constantine to Julian: Pagan and Byzantine Views, A Source History* (Routledge, 1996).

On many coins, we see the Undefeated Sun present to Constantine the cosmic sphere marked with an X – Plato's celestial intersection, the Gates of Heaven that the emperor controls.

This heavenly mandate was given a Christian retooling by the bishop Lactantius, which describes the dream before the battle at the Milvian Bridge. Only one copy of this text has survived – a medieval document found in a Benedictine monastery – which has raised eyebrows about its authenticity.

> Constantine was advised in a dream to mark the heavenly sign of God on the shields of his soldiers and then engage in battle. He did as he was commanded and by means of a slanted letter X with the top of its head bent around, he marked Christ on their shields. Armed with this sign, the army took up its weapons. – Lactantius, *De Mortibus Persecutorum*

We should note that here we have just a dream, and not a wide-screen vision in the heavens that everyone could see. Notice also that the "heavenly sign" is the letter X – Plato's cosmic X that Justin Martyr had earlier linked to the Son of God. By the mere bending around of one end of Plato's X... Voila! It miraculously becomes the mark of Christ.

After Constantine died, the bishop Eusebius of Caesarea would claim that the emperor had given him confidential information under oath, and he painted quite a different 'vision.'

> About the time of the midday sun, when day was just turning, he said he saw with his own eyes, up in the sky and resting over the sun, a cross-shaped trophy formed from light, and a text attached to it which said, 'By this conquer'. Amazement at the spectacle seized both him and the whole company of soldiers which was then accompanying him on a campaign he was conducting somewhere, and witnessed the miracle. – Eusebius, *Life of Constantine*

Amazingly, we now have a visible noonday spectacle that is cross-shaped, and furthermore, we see letters in the heavens that spell out a legible message. This early example of skywriting propaganda shows that the text of Eusebius is pure fiction, a campaign to wrest control of the Gates of Heaven from under Plato's authority – the cosmic X – in order to re-brand it as a cross-shaped symbol controlled by the Christian hierarchy.

Coins of Constantius II illustrate that the propaganda of Eusebius was swallowed whole for political consumption: the motto "Hoc Signo Victor Eris" surrounds the labarum sporting the Chi Rho, while the pagan symbols of the planets along the standard have been erased (Figure 6).

Figure 6. Coin of Constantius II with the emperor carrying the labarum with Chi Rho under the legend 'Hoc Signo Victor Eris' (This sign is the victor) – the planetary symbols along the shaft have disappeared.

Christians crowed that the Edict of Milan had granted them religious freedom, yet they themselves did not live up to the precepts that tolerated all faiths. As soon as they gained power, they wiped out all traces within the Empire of pagan creeds, of Manichaeism, of competing Christian views held by Arians, Gnostics, and other 'heretics.'

Conclusion

The texts of the bishops Lactantius and Eusebius give vastly different accounts of Constantine's so-called vision, making them both suspect. Eusebius makes no mention of the battle at the Milvian Bridge, saying instead that Constantine was off campaigning somewhere when he and his troops witnessed a celestial diorama with visible letters spelling out a message of victory to the emperor.

The quiet dream in Lactantius' supposed account does not match the daytime spectacle that Eusebius claims Constantine revealed to him. The many oaths of Constantine that Eusebius repeatedly flogs show that the bishop 'doth protest too much' and that his tale is pure invention.

All these discrepancies were ignored for the sake of a hybrid confabulation that persists to this day. Supposedly, before the clash at the Milvian Bridge, Constantine and his troops witnessed a sign in the heavens in broad daylight. But according to the writings of the bishops, such an event never happened.

What did happen was the re-shaping of an ancient symbol that stood for the heavenly gates – Plato's X, the letter Chi – into a Christian symbol by the addition of a vertical intersect, giving us the Chi Rho that can be found in many churches today.

Not a problem, said many pagans at first. We climb up the planetary ladder and we arrive at the gates of heaven that are promised to us by the intersecting celestial sign, whether that is Plato's Chi or the Christian Chi Rho. But the Christians would play a mean trick on the unsuspecting opposition.

They kicked out the ladder from under the pagans. The Planets would be demonized (Lucifer) where once they had been divine (Diana Lucifera).

Yet the influence the Wanderers once exerted on the lives

of our ancestors is still felt today. The names of the days of the week can be traced to the Wanderers in the sky (Saturnday, Sunday, Moonday, etc.).

In 321 AD, Constantine made the planetary seven-day week the official timekeeper of the Empire, enshrining the circular dance of the celestial bodies into our memes for centuries to come.

Like the army standards with planetary symbols topped by the Chi Rho, the planetary week with its climax in the day of the Sun reveals Constantine's attempt at fusing the ancient astral religion and culture of Rome with the newly emerging Christian cult. That was the true vision of Constantine.

Figure 7. Signed pewter medal of 1913 showing the daytime apparition in the heavens that, according to the writings of the bishops, never happened.

Plato's Cosmic X

HEAVENLY GATES AT THE CELESTIAL CROSSROADS

PROCEEDINGS OF 2012 SEAC (SOCIETE EUROPEENNE POUR L'ASTRONOMIE DANS LA CULTURE) CONFERENCE, LJUBLJANA, SLOVENIA

"THIS IS NOT THE WAY THINGS were supposed to go," thought Cato.

Didn't Rome's greatest general, Pompey, agree to champion the cause of the Republic? Didn't a large army assemble under Pompey's command in Greece to face the looming threat of dictatorship?

Yet the upstart Julius Caesar had triumphed over Pompey, hunting him all the way to Egypt, where Cleopatra lay in wait like a spider. The defenders of the Republic managed to regroup, but Caesar would pursue them relentlessly across North Africa.

Upon receiving news that Caesar had defeated the last troops that stood in his way, Cato put his affairs in order, retired to his chambers to read Plato's discourse on the immortality of the soul in *Phaedo,* and then, like a true Roman, he took his own life.

Plato's writings – with the immortality of the soul – were the Bible of the Greco-Roman world over centuries, laying out a cosmology and soteriology that promised a blessed afterlife to virtuous souls.

Plato's Heavenly Gates

At some point, we've all heard jokes that go something like this: "Three men die and they arrive at the heavenly gates guarded by Saint Peter, who asks each man a probing question..."

Now, where did the heavenly gates originate?

It might come as a surprise to some that the celestial portals already appear in Plato's *Republic*, written around 350 BC. At the end of that tome, in the story of the fallen warrior Er, Plato describes heavenly gates through which the souls of the just ascend to the celestial abode.

> There were two openings in the earth, and above them two others in the heavens, and between them judges sat. These, having rendered their judgment, ordered the just to go upwards into the heavens through the door on the right...– Plato, *Republic*[1]

Plato also influenced the great Roman statesman Cicero who, like Cato, would lose his life for defending the Republic. Over the course of his literary career, Cicero would translate Plato's *Timaeus* and emulate his other important works (*Laws, Republic*). The 'Dream of Scipio' at the end of Cicero's *On The Republic* is in fact a reworking for a Roman audience of the 'Myth of Er' at the end of Plato's *Republic*.

Cicero's protagonist meets his adoptive father and grandfather in the Milky Way, the abode of virtuous souls. Then Scipio hears the Music of the Spheres, the harmony generated by the movement of the Planets along the ecliptic, the path that traces out the Zodiac.

These two intersecting paths would be of utmost importance, as they point to a specific location in the heavens, a location evoked by an unplaced fragment from Cicero's *On The Republic*.

> 'If it is right for any man to climb to the tracts of the heavenly ones, then the great gate of heaven lies open for me alone... It is so indeed, Africanus, that same door lay open also to Hercules.' – Lactantius, *Divine Institutes*[2]

The "tracts of the heavenly ones" is of course the course of the planetary gods in their celestial chariots, the ecliptic that runs through the constellations of the Zodiac. Somewhere along this path, somewhere along the Zodiac, lies "the great gate of heaven" that is invoked by Cicero, who was following his hero, Plato, who had already given us the celestial portals.

The heavenly gates could be reached by heroic strivings, such as the Labors of Hercules that were often shoehorned into a dozen, to echo the twelve astrological signs through which the heavenly Wanderers traveled. Where along the planetary path of the heroes and the gods can we find the gates of heaven?

Macrobius' Heavenly Gates at the Intersections

Four centuries after the demise of the Republic, the Roman writer Macrobius employed Cicero's 'Dream of Scipio' as a framework for his understanding of Platonist cosmology, and to reveal that the Gates of Heaven stand at the intersections in the firmament.

> The Milky Way girdles the zodiac, its great circle meeting it obliquely... Souls are believed to pass through these portals when going from the sky to the earth and returning from the earth to the sky... – Macrobius, *Commentary on the Dream of Scipio*[3]

Macrobius links us back in time directly to Cicero, where we find the Milky Way and hear the music of the planetary spheres. These elements can already be found in Plato's 'Myth of Er,' where celestial gates lead to a 'pillar of light' – the Milky Way – and the harmony of the planetary Sirens points to the movement of the Planets along the ecliptic that traces out the Zodiac.

From Macrobius to Cicero to Plato – the path of Platonist cosmology can be traced back cleanly across seven centuries. The two intersecting structures in the sky are the Milky Way and the path of the Planets along the Zodiac, and Macrobius squarely places the celestial portals at these intersections.

Plato's Heavenly Intersections

The celestial intersections can be found in Plato's *Timaeus*, where the Creator fashions a long strip of cosmic material that he splits in half lengthwise. He places these at an angle to each other, like an X, and then he bends the extremities in a curve until they meet each other on the opposite side of a sphere – giving two intersecting circles on the celestial sphere, one the circle of the Same, the other the circle of the Different..

> Next, he sliced this entire compound in two along its length, joined the two halves together center to center like an X, and bent them back in a circle, attaching each half to itself end to end and to the ends of the other half at the point opposite to the one where they had been joined together. He then included them in that motion which revolves in the same place without variation, and began to make one the outer, and the other the inner circle. And he decreed that the outer movement should be the movement of the Same, while the inner one should be that of the Different. – Plato, *Timaeus*[4]

In order to gives us a clue, Plato reveals that one of these circles is the path of the Wanderers, the Planets that travel along the ecliptic.

> When the god had finished making a body for each of them, he placed them into the orbits traced by the period of the Different – seven bodies in seven orbits. – Plato, *Timaeus*[5]

In Plato's own words, we clearly find celestial portals through which virtuous souls ascend to the divine abode (*Republic*) and two circles that intersect in the heavens (*Timaeus*), one of them being the course of the Planets.

Visible Intersections in the Heavens

The crossroads in the sky can be found in Manilius' *Astronomica* from the time of Augustus, where the author, a great admirer of Plato, claims that this phenomenon is visible in the night sky.

> To these [previous circles] you must add two circles which lie athwart and trace lines that cross each other. One contains the shining signs

through which the Sun plies his reins, followed by the wandering Moon in her chariot, and wherein the five planets that struggle against the opposite movement of the sky perform the dances of their orbits...

Nor does it elude the sight of the eye, as if it were a circle to be comprehended by the mind alone, even as the previous circles are perceived by the mind: nay, throughout its mighty circuit it shines like a baldric studded with stars and gives brilliance to heaven with its broad outline standing out in sharp relief.

The other circle [the Milky Way] is placed crosswise to it...

– Manilius, *Astronomica*[6]

The only time the path of the planets "gives brilliance to heaven with its broad outline," as Manilius claims, is when the zodiacal light, a broad swath of interplanetary dust, envelops the Wanderers, a rare sight best seen in the fall and in the spring.

And when the zodiacal light intersects the Milky Way (Figure 1), we witness the cosmic X that Plato describes in *Timaeus*, the intersection that Macrobius calls the celestial gates, the heavenly portals that Plato describes in the Myth of Er at the end of *Republic*.

Figure 1. THE ZODIACAL LIGHT rises from the horizon, envelops planets along the ecliptic, and intersects the Milky Way, revealing the celestial crossroads – Plato's X – the location of the Gates of Heaven. (Photo: Matt BenDaniel)

Celestial Intersection on Roman Coins

The celestial intersection, Plato's X, could be seen on Imperial coins over hundreds of years, where it proclaimed Rome's control of the Gates of Heaven (Figure 2). The cosmic crossroads would appear on coins of the emperors Domitian, Antoninus Pius, Marcus Aurelius, Lucius Verus, Macrinus, Licinius, Constantine, Constantine's sons, etc.

Figure 2. PLATO'S X ON ROMAN COINS. Left: Denarius of Antoninus Pius, with Italia enthroned on celestial sphere with intersecting lines (RIC III, 98C). Middle: Coin of Constantine, with Jupiter presenting the cosmic orb with intersecting lines and dotted with stars (RIC VI, Cyzicus 80). Right: Coin of Constantine's son, Constans, where he grasps the heavenly globe with stars and intersecting lines (RIC VIII, Rome 158).

We should not think that the X in the sky was a latecomer to Roman cosmology because it could already be found on coins of the Republic. A denarius struck by Manlius (c.107 BC) shows in the center the Sun god in his quadriga, on the right the crescent Moon, and on the left the X in the sky, the intersection that indicates the heavenly gates (Figure 3).

Figure 3. DENARIUS OF MANLIUS shows the Sun in quadriga, Moon on the right, and the celestial X on the left.

Wonderfully depicted on the coin of Manlius, we see the path of the Wanderers – the Sun and the Moon – as well as the celestial X – Plato's cosmic X – the intersection in the sky that stood for the Gates of Heaven.

Conclusion

Although most Christians today do not know that heaven already existed in the pagan mind, a mechanism for reaching the celestial abode was in place long before the Christian era. Hundreds of years before the Christian cult made its appearance, Plato described heavenly gates through which virtuous souls entered a blissful afterlife.

Erased from our cultural memory for more than a thousand years, the pagan Gates of Heaven at the celestial intersections offered a blissful afterlife in the Milky Way to those who followed the ways of the planetary gods and goddesses.

Unfortunately with the expansion of light pollution, fewer and fewer will ever see the Milky Way and even fewer will ever witness the Zodiacal Light, the two visible circles in the night sky that revealed Plato's cosmic X, the crossroads that once indicated the Gates of Heaven.

Notes

[1] Cooper, *Plato: Complete Works*, p. 1218.

[2] Bowen, Garnsey, *Lactantius: Divine Institutes*, p. 101

[3] Stahl, *Macrobius: Commentary On the Dream of Scipio*, p. 133.

[4] Zeyl, *Plato: Timaeus*, p. 21.

[5] Zeyl, *Plato: Timaeus*, p. 25.

[6] Goold, *Manilius: Astronomica*, p. 57-59.

References

Bowen, Anthony and Garnsey, Peter (trans.), *Lactantius: Divine Institutes*. Liverpool University Press, 20XX.

Cooper, J. M. (ed), *Plato: Complete Works*. Hackett Publishing, 1997.

Gould, G. P. (trans.), *Manilius: Astronomica*. Harvard University Press, 1977.

Stahl, William Harris (trans.), *Macrobius: Commentary on the Dream of Scipio*. Columbia University Press, New York, 1990.

Zeyl, J. Donald, *Plato: Timaeus*. Hackett Publishing. Indianapolis / Cambridge, 2000.

Plato's Visible God

THE COSMIC SOUL REFLECTED IN THE HEAVENS

RELIGIONS online peer-reviewed journal, September 2012

IN AN OPEN LETTER ADDRESSED to the Roman emperor Antoninus Pius, the Christian apologist Justin Martyr argued that the celestial X in Plato's Timaeus was a foreshadowing of the Christian cross, a shape that Plato had somehow pilfered from Moses.

> And the scientific discussion of the Son of God in his *Timaeus* – when he says: "He arranged him as an X in the whole" – Plato took from Moses, and spoke in similar terms. – Justin Martyr, *Apology on Behalf of Christians* ([1], p. 235)

Here Justin conflates the Christian Son of God with Plato's Cosmic Soul, and he provides evidence that, in the glory days of the Roman Empire, Plato's Anima Mundi was seen as composed of intersecting lines in the heavens.

At the twilight of the Empire, the Platonic successor Proclus likewise connected Plato's intersecting symbol with the cosmic realm.

> The shape X itself that results from the affixing [of the two strips] has the highest degree of appropriateness to the universe and to the soul. – Proclus, *Commentary on Plato's* Timaeus ([2], p. 233)

The X in the universe that Proclus is commenting on and the Son of God that Justin Martyr places like an X in the cosmos both refer to the passage in Plato's Timaeus that describes the creation of the World Soul.

> Next, he sliced this entire compound in two along its length, joined the two halves together center to center like an X, and bent them back in a circle, attaching each half to itself end to end and to the ends of the other half at the point opposite to the one where they had been joined together. He then included them in that motion which revolves in the same place without variation, and began to make one the outer, and the other the inner circle. And he decreed that the outer movement should be the movement of the Same, while the inner one should be that of the Different. – Plato, *Timaeus* ([3], p. 21)

Here Plato positions two cosmic circles at an angle to each other so that they intersect "like an X." What might these two circles stand for?

Modern and Ancient Mis-Interpretation

One interpretation, put forth by R. G. Bury in 1929 and Francis Cornford in 1937, maintains that Plato's celestial X is formed by the intersection of the celestial/sidereal equator and the ecliptic/zodiac.

> He now tilts the inner band, so that it makes an oblique angle with the outer, which is set at the horizontal; from which we see that the Revolution of the Same represents the celestial Equator, moving "horizontally to the right" (from East to West), and the Revolution of the Other represents the Ecliptic, which moves in a contrary direction to the Equator (from West to East), and at an angle to it. The Ecliptic He divides into seven, to represent the seven planets. – Bury, Plato: Timaeus, ([4], p. 72, n. 1)

> Timaeus now speaks as if the Demiurge had made a long band of soul-stuff, marked off by the intervals of his scale. This he proceeds to slit lengthwise into two strips, which he puts together by their middles and bends round into two circles or rings, corresponding to the sidereal equator and the Zodiac. – Cornford, Plato's Cosmology ([5], p. 72)

Plato makes no mention of the celestial equator, yet influential books on the cult of Mithras in the Roman Empire were released by David Ulansey [6] and Roger Beck [7], with theBury/Cornford interpretation championed by both

Ulansey[1] and Beck.[2]

Where might Bury and Cornford have found this interpretation of Plato's X? Perhaps in the works of Proclus, whose Commentary on Plato's Timaeus [2] already espoused this view.

> Now surely two circles come into being, and these have come to be in such a way that one is on the inside and the other is on the outside, and they are at an angle to one another. Now one of these is called the circle of the Same and the other is the circle the Different. The one corresponds to the equator while the other corresponds to the circle of the ecliptic. The entire circle of the Different is carried around the ecliptic, while the circle of the Same is carried around the equator. Because of this it is immediately evident that it is not necessary to assume these circles to be at right angles to one another, but rather like an X, just as Plato said... – Proclus, Commentary on Plato's Timaeus ([2], p. 222).

Proclus' equator and Cornford's sidereal equator refer to the celestial equator, a projection of the Earth's equator into the heavens, showing that Proclus and Bury and Cornford (and Ulansey and Beck) all agree that Plato's X is composed of the path of the planets (the ecliptic that traces out the zodiac) and the celestial equator.

But Plato's own words prove this view to be in error.

At the end of Timaeus, Plato proclaims that the living Cosmos (which he said had the shape of an X) is a visible, discernible god.

> And so now we may say that our account of the universe has reached its conclusion. This world of ours has received and teems with living things, mortal and immortal. A visible living thing containing visible ones, perceptible god, image of the intelligible Living Thing...Our one heaven, indeed the only one of its kind, has come to be. – Plato, Timaeus, ([3], p. 88)

But since the celestial equator is a mathematical calculation and a geometric projection, it is certainly not visible in the heavens – exposing the fallacy behind the interpretation of Proclus, Bury, Cornford, Ulansey, Beck, etc.

Correct Interpretation

Who should we look to for the correct explanation? To Plato himself, whose narrative emphasizes that the Cosmic Soul, which has the shape of an X, will find a visible manifestation in the heavens.

The Demiurge first creates the World Soul, composed of two cosmic circles whose intersections take the form of an X. According to Plato, the Cosmic Soul itself is invisible, but the Creator decides to make the visible body of the universe as similar as possible to the invisible soul.

> Now while the body of the universe had come to be as a visible thing, the soul was invisible. – Plato, Timaeus, ([3], p. 23)

> Now when the Father who had begotten the universe observed it set in motion and alive, a thing that had come to be as a shrine for the everlasting gods, he was well pleased, and in his delight he thought of making it more like its model still. So, as the model was itself an everlasting Living Thing, he set himself to bringing this universe to completion in such a way that it, too, would have that character to the extent that was possible. – Plato, *Timaeus*. ([3], p. 23-24)

In other words, the visible body of the universe mirrors as closely as possible the form of the invisible Cosmic Soul, whose two intersecting circles give the shape of an X. To give us a clue as to the components of this cosmic scheme, Plato reveals that one of the celestial circles, the circle of the Different, follows the path of the Wanderers, the seven Planets that trace out the zodiac.

> When the god had finished making a body for each of them, he placed them into the orbits traced by the period of the Different – seven bodies in seven orbits. – Plato, *Timaeus*, ([3], p. 25)

But does the course of the Planets ever become visible, as Plato stipulates?

And what other visible circle in the sky intersects the planetary path to form an X, the shape of the Cosmic Soul that is mirrored in the heavens?

We can thank our lucky stars that the Roman writer Manilius penned an exposition of the art of astrology around the time of Augustus. At the beginning of his esoteric tome, Manilius emphatically describes two visible circles that intersect in the heavens. One is the path of the Planets and the other is the Milky Way.

> To these [previous circles] you must add two circles which lie athwart and trace lines that cross each other. One contains the shining signs through which the Sun plies his reins, followed by the wandering Moon in her chariot, and wherein the five planets which struggle against the opposite movement of the sky perform the dances of their orbits that nature's law diversifies... Nor does it elude the sight of the eye, as if it were a circle to be comprehended by the mind alone, even as the previous circles are perceived by the mind: nay, throughout its mighty circuit it shines like a baldric studded with stars and gives brilliance to heaven with its broad outline standing out in sharp relief.
>
> The other circle [the Milky Way] is placed crosswise to it. – Manilius, Astronomica ([8], p. 57, 59)

Certainly the Via Galactica is visible even today at a good distance from light-polluting cities, and since this awesome apparition in the night sky partakes of the revolution of the fixed stars – Plato's motion of the Same – it becomes evident that the Milky Way is the component of Plato's visible celestial X that intersects the path of the Planets.

What is the other celestial apparition that is as visible as Manilius claims?

Zodiacal Light & Milky Way

The Wanderers along the ecliptic map out the constellations of the Zodiac, but neither the planets themselves nor the zodiacal constellations give the "brilliance to heaven... standing out in sharp relief" that Manilius paints before our eyes. What celestial phenomenon might explain Manilius' exuberant evocation?

That would be a rare and miraculous-seeming event that occurs only at specific times of the year. In temperate zones, the zodiacal light illuminates the sky along the ecliptic shortly before dawn or soon after dusk, depending on the season.

Enveloping the planets along its path, this broad swath of interplanetary dust reveals the stairway to heaven along which the souls of the just climb to the Milky Way, the celestial abode according to Cicero (Dream of Scipio), Manilius (Astronomica), Ovid (Metamorphoses), Macrobius (Commentary on the Dream of Scipio), Martianus Capella (Marriage of Philology and Mercury), etc.

And when the heavenly intersection reveals itself at rare times of the year, we witness Plato's visible, perceptible god – the embodiment of the Cosmic Soul – that, according to Justin Martyr, looks like "an X" (Figure 1).

Figure 1. The zodiacal light rises from the horizon, envelops planets along the ecliptic, and intersects the Milky Way, revealing Plato's visible god, the celestial X that mirrors the World Soul (Photo: Matt BenDaniel).

Plato's perceptible god, the visible intersection in the sky, appeared on coins of the Roman Empire for hundreds of years (Figure 2), on coins minted by Antoninus Pius, Marcus Aurelius, Macrinus, and other emperors.

Figure 2. Plato's X on Roman coins. Left: Denarius of Antoninus Pius, with Italia enthroned on celestial sphere with intersecting lines (RIC III [9] #98a). Middle: Coin of Marcus Aurelius, with Providentia pointing to the celestial sphere with intersecting lines (*RIC* III [9] (Pius) #446). Right: Coin of Macrinus, with Providentia pointing to cosmic orb with intersecting lines (*RIC* IVPt 2 [10] #80).

In his plea to the Roman emperor, Justin Martyr gives us not only Plato's *visible god* placed "as an X" in the heavens, as shown on imperial coins, he also testifies to a most important use of the intersecting symbol in the Roman world.

> And you set up the images of your dead emperors on this pattern, and you name them gods through inscriptions. – Justin Martyr, Apology on Behalf of Christians ([1], p. 227)

Invoking the sacred intersection struck on imperial coins of his day, Justin points to a crucial ritual in Roman religious life, one that harkened back to the funeral of Augustus himself. When a beloved emperor died, an eagle was released as the pyre consumed the physical body, to portray the soul of the emperor being carried by the bird of Jupiter to the company of the gods.

The Roman Senate deified admired rulers with the title *divus*, or divine, in the process of *consecratio* (Figure 3). The cosmic orb, with the intersecting lines that Justin Martyr refers to in his public letter, stood for the World Soul that

29

encompassed the heavens where the emperor would enjoy eternal life among the gods.

Figure 3. *Consecratio* denarius declares Marcus Aurelius a divine being (*divus*), while Jupiter's eagle sits atop the celestial orb with intersecting lines: Plato's X, the visible reflection of the Cosmic Soul in the heavens (RIC III [9] (Commodus) #273)

Conclusions

With Justin Martyr's testimony, we can connect the *consecratio* coins of Roman emperors to the celestial intersecting symbol that indicates Plato's perceptible god. In this symbolism, we see the soul of the emperor joining the gods in the heavens, with the visible X indicating Plato's Cosmic Soul.

More than a hundred years after Justin Martyr's revisionist attempt, the Christian bishop Lactantius again sought to recast Plato's celestial X into a Christian symbol.[3] Following the lead of Lactantius, the bishop Eusebius of Caesaria conjured a masterpiece of propaganda[4] that would eclipse the memory of Plato's visible cosmic intersection and erase it from the Western mind for more than a millennium and a half.

The tradition of distorting Plato's celestial intersection into a Christian symbol endures in Bury's translation of Timaeus (1929), where Plato's X is mistranslated as "like a great cross" ([4], p. 71).

Thanks to the *Astronomica* of Manilius, however, we can be sure that Plato's celestial X was composed of the intersection of two visible celestial circles: the Milky Way and the zodiacal light that illuminates the path of the Planets, the two perceptible structures in the heavens that embody Plato's Visible God.

Notes

[1] "The celestial equator and the ecliptic intersect at two points..." (Ulansey, p. 47)

[2] "The model... is essentially that established by Plato in the *Timaeus*... It even has its standard iconographic representation: the world globe with the crossed bands of equator and ecliptic/zodiac." (Beck, p. 79)

[3] "Constantine was advised in a dream to mark the heavenly sign of God on the shields of his soldiers and then engage in battle. He did as he was commanded and by means of a slanted letter X with the top of its head bent around, he marked Christ on their shields. Armed with this sign, the army took up its weapons." – Lactantius, *De Mortibus Persecutorum* ([11], p. 63)

[4] "About the time of the midday sun, when day was just turning, he said he saw with his own eyes, up in the sky and resting over the sun, a cross-

shaped trophy formed from light, and a text attached to it which said, 'By this conquer'." – Eusebius, *Life of Constantine* ([12], p. 81)

References

[1] D. Minns and P. Parvis, eds. *Justin, Philosopher and Martyr: Apologies*. Oxford University Press, New York, NY, USA, 2009.

[2] D. Baltzly, trans. *Proclus: Commentary on Plato's Timaeus: Volume 4, Book 3, Part 2, Proclus on the World Soul*. Cambridge University Press, New York, NY, 2009.

[3] D. J. Zeyl, trans. *Plato: Timaeus*. Hackett Publishing Company, Indianapolis, IN, USA, 2000.

[4] R. G. Bury, trans. *Plato: Timaeus, Critias, Cleitophon, Menexenus, Epistles*. Harvard University Press, Cambridge, MA, USA, 1929.

[5] F. M. Cornford. *Plato's Cosmology: The Timaeus of Plato*. K. Paul, Trench, Trubner, London, UK / Harcourt, Brace, New York, NY, USA, 1937.

[6] D. Ulansey. *The Origins of the Mithraic Mysteries*. Oxford University Press, New York, NY, USA, 1989.

[7] R. Beck. *The Religion of the Mithras Cult in the Roman Empire*. Oxford University Press, New York, NY, USA, 2006.

[8] G. P. Goold, trans. *Manilius: Astronomica*. Harvard University Press, Cambridge, MA, USA, 1977.

[9] H. Mattingly and E. A. Sydenham. *The Roman Imperial Coinage: Antoninus Pius to Commodius*, Vol. III. Spink & Son, London, UK, 1930.

[10] H. Mattingly, E. A. Sydenham, and C. H. V. Sutherland. *The Roman Imperial Coinage: Macrinus to Pupienus*. Vol. IV, Part 2. Spink & Son, London, UK, 1938.

[11] J. L. Creed. *Lactantius: De Mortibus Persecutorum*. Oxford University Press, New York, NY, USA, 1984.

[12] A. Cameron and S. G. Hall, trans. *Eusebius: Life of Constantine*. Oxford University Press, New York, NY, USA, 1999.

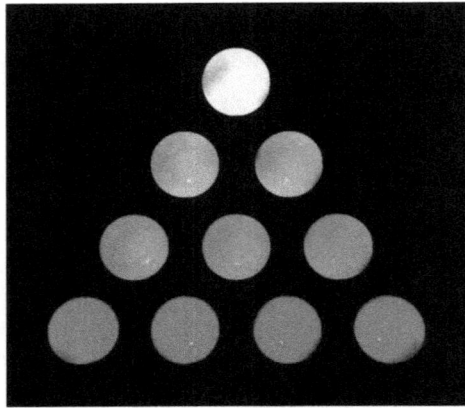

To Read God's Mind

Pythagoreans & The Birth of Science

Parabola quarterly journal, Winter 2012

"THIS IS NOT THE WAY THINGS WERE supposed to go,"
thought Plato. As he squatted in the fetid belly of a cargo ship,
iron chains rubbed his wrists and ankles raw.

Just a few weeks ago, he had been warmly greeted by
leading citizens of Tarentum and ot`her rich cities of Magna
Graecia. Heated discussions and drinking bouts long into the
night, leisurely walks in lush orchards seemed to cement
friendships with cultivated Greeks from the colonies, whose
sons brought back trophies from the Isthmian Games while
their horses triumphed at the Olympiads.

What had derailed those halcyon days? Hadn't Archytas
welcomed Plato into his home? Archytas, who had been

proclaimed *strategos* seven times by the citizens of Tarentum even though the city laws forbade any man from serving as general more than once, from fear of power going to his head? Wasn't Archytas one of the most prominent Pythagoreans of the age, who reportedly solved the puzzle posed by the Oracle of Delphi ("Double the altar of Apollo – a cube"), thus saving Delos from the plague?

But then had come Dion, from the island of Sicily, singing the praises of Dionysius, ruler of the city of Syracuse. Dion promised that Dionysius would heap honors on the philosopher from Athens and would eagerly listen to his every word. But the young tyrant had a short attention span, and at a banquet, he called Plato a doddering old fool. Plato retorted that he was but an uneducated bully. Mutual friends separated the two, and at first it looked as if matters had been settled amicably. But the seeds of enmity had found root, and Dionysius conspired with the ambassador from Sparta. Traveling on the same ship as Plato, once at sea, the Spartan had Plato cast into chains, to be sold into slavery.

As soon as the Pythagoreans heard of this treachery, they dispatched a rescue party to buy Plato's freedom. As a philosopher, Plato was greatly animated by the notion of Justice, and his sense of justice would surely dictate that he owed the Pythagoreans not just coins, but deep gratitude. Yet in Plato's voluminous works, across dozens of dialogues, we find the Pythagoreans mentioned just a handful of times.

This seeming ingratitude points to a little-known facet of a complicated relationship: Plato had obviously been initiated into the Pythagorean brotherhood, and members who revealed the secret teachings were ostracized and banished from memory. All Plato could do was invent myths and bury the secrets in there.

Later generations would point a finger and accuse: "Plato Pythagorizes...!" Already Aristotle, Plato's own disciple, lumped Plato with the Pythagoreans.

> The Pythagoreans say that things exist by 'imitation' of numbers, but Plato, by 'participation,' but these are the same, and Plato only

Everything is Number. That was the great leap of the Pythagoreans, that everything could be measured and quantified. That was the birth of science, where repeated experiments yield predictably measurable results.

Another link to the brotherhood can be found in Plato's most Pythagorean text, *Timaeus*, where the Demiurge creates the cosmos according to ratio and number. When describing the planetary system, Plato invokes the celestial music of the Pythagoreans.

> Up above the rim of each circle sat a Siren, singing one pure note. And the music of the different spheres gave one harmony. – Plato, *Timaeus*

The harmony of the Sirens could be found in a Pythagorean *akousma*, a koan-like riddle that only those from the inner circle could explain.

Q: What is the Oracle at Delphi?

A: The Tetraktys that gives the Harmony of the Sirens

The Tripod of the Oracle of Delphi on an incuse silver stater of Croton, from the lifetime of Pythagoras.

The Oracle of Delphi is the oracle of Apollo, the Sun god whose seven-string lyre leads the dance of the celestial Wanderers, the seven Planets. What is the Harmony of the Sirens? It's the celestial music the Planets produce as they revolve in their orbits.

How does the Tetraktys link these diverse elements together? When the Pythagoreans had to swear an oath, they would call upon this pyramid of pebbles, one at the top, two in the next row, three in the third, and four in the bottom row.

Pythagorean Oath: "By he who gave our souls the Tetraktys."

The true beauty of the Tetraktys jumps out when the whole-number ratios embedded in it reveal the foundations of Western music, an encyclopedia written with 10 little stones.

Pythagoras portrayed in Raphael's School of Athens, in the Vatican. By his left foot, a chalkboard shows the Tetraktys.

The two top rows, the singleton and the next row of two pebbles, give the ratio of 1:2, the musical interval of the Octave, the universal basis of music. The two middle rows give the ratio of 2:3, the musical interval of the Fifth, the next most consonant interval of the musical scale. And the bottom two rows spit out the ratio of 3:4, the musical interval of the Fourth, completing the canon and providing all that's necessary to develop a mathematical theory of music.

If you add the interval of the Fourth to the interval of the

Fifth, you get a perfect Octave, the Tetraktys itself. If you subtract the Fourth from the Fifth, you get a whole note, the basic unit of the musical scale. According to the myth, Pythagoras discovered these musical intervals when he walked by the ringing shop of a smith, whose pounding hammers beat out different tones on metal bars of varying length.

Pythagoras moved on to string theory, building a monochord that showed that filaments of differing lengths gave intervals of mathematically predictable notes, with the pitch measurably related to the length of the string. In his investigations, Pythagoras came upon an astonishing fact that still bears his name.

No, it's not the Pythagorean Theorem, which every schoolchild hears described these days, but with little elucidation about Pythagoras. Rarely is it mentioned that Plato proclaimed his allegiance to the Pythagoreans with the motto engraved at the entrance to his Academy.

"Let no one ignorant of Geometry enter here."

Yes, the Pythagoreans and Plato's Academy both demanded a basic grasp of logic, which was best provided by geometry and mathematics. Numbers were held in such high regard that they sometimes acquired almost mystical properties.

"Seven is the key to the universe," wrote Cicero in the Dream of Scipio, and he dedicated his translation of Plato's Timaeus to his Pythagorean mentor, Nigidius Figulus. Divide the number 1 by the numerals from 2 to 9, if you wish to divine Cicero's secret — even Isaac Newton wouldn't let go of this esoteric viewpoint.

But it was music and astronomy that were reserved for the highest echelons of investigation. In astronomy, the seven Wanderers mirrored the seven notes of the musical Octave, giving the celestial harmony of the Sirens and suggesting that these widely different branches of learning were in fact sister sciences.

In music, Pythagoras noted that a cycle of seven octaves matches a cycle of twelve fifths almost exactly. Almost, but not exactly... The small difference is called the Comma of Pythagoras, a seal that proves the presence of the Pythagoreans at the birth of modern science.

South Italian wine-drinking kylix with a Pythagorean motif of seven leaves. Cicero: "Seven is the key to the universe."

The cycling of fifths through octaves produces each and every halftone of the musical scale, the 12 halftones marked out by the 12 fifths that revolve through seven octaves. In his book on music theory, Michael Pilhofer (MM), writes:

> The creation and use of the Circle of Fifths is the very foundation of modern Western music theory...

Not just the foundation of modern music theory, but music theory going all the way back to Pythagoras, who stands at the beginning of the long corridor of time, holding a torch to illumine this dark path.

So important was the cycling of fifths through octaves to the Pythagoreans that they encoded it into something that people use every day, yet whose origins remain shrouded in oblivion. Each day, as you wake up, bring to mind the days of the Week that reflect the musical Circle of Fifths that cycles through the Octave, the seven days that are named after the seven Planets.

Moon day (Monday) is the first step up the planetary ladder, as told by Plato, Vitruvius, Martianus Capella, etc. From there, an interval of a fifth (skipping three Planets) lands us at Mars, the French Mardi. Another fifth circling around brings us to Mercredi, the day of Mercury. A further leap gives us Jeudi, or Jove/Jupiter day, and then we arrive at Vendredi, the day of Venus. One more hop brings us to Saturn (Saturday), and finally we rest on Sunday. The next day, we start the week all over again, pushing the boulder up the hill, just like Sisyphus.

The Pythagorean Music of the Spheres was the grand unifying theory of the ancient world, and Pythagorean thought would inspire those who opened the floodgates of modern science thousands of years later.

In *Standing On The Shoulders of Giants*, physicist and mathematician Stephen Hawking compiled the scientific works that most impacted Western scientific thinking. The authors he chose: Copernicus, Galileo, Kepler, Newton, and Einstein.

Who inspired the revolutionary vision of the heliocentric planetary system? According to Copernicus, it was Hicetas and Ecphantus of Syracuse, Pythagorean scholars from Sicily.

Johannes Kepler would be bounced by fortune from posts as a simple mathematics professor to highly regarded astronomical scientist. Tycho Brahe hired him as his lieutenant, to aid in the meticulous observations of the movements of the planets. When Brahe died, Kepler got his hands on jealously guarded data that had tracked the motion of Mars for decades, providing concrete numerical evidence for Kepler's Third Law of Planetary Motion.

Where did Kepler look for inspiration for his theories of planetary motion? First he tried the geometric model of the Platonic solids, but that did not prove entirely tractable. Digging further back to the Pythagoreans, Kepler found the

Harmony of the Sirens that would yield his masterpiece, Harmonice Mundi, where Kepler still describes the orbits of the Planets in musical terms.

Hoping for assistance from the spirit of Pythagoras, Kepler mused that perhaps "the soul of Pythagoras has migrated into me." Through the prism of Pythagorean Number, Kepler laid the foundations of the Space Age, and his Laws of Planetary Motion are still used today to calculate the trajectory of rockets and interplanetary satellites.

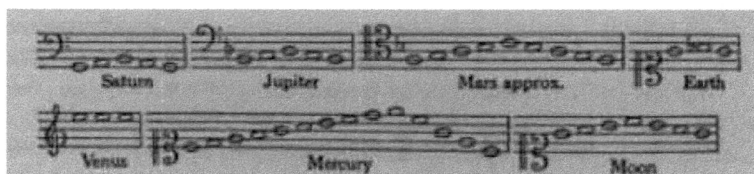

Kepler's Harmony of the Planets in *Harmonice Mundi*, from Stephen Hawking's *On The Shoulders of Giants*

Isaac Newton would build on Kepler's work to march us into the modern age, but even he retained a fondness for the symmetry inherent in Pythagorean thought. With a glass prism, he broke up a ray of white light into its components: red, orange, yellow, green, blue, and violet. But these are only six colors. For order to be maintained, Newton had to insert a seventh color. That's how we got indigo, a color that does not exist independently in the spectrum, but which was necessary, in Newton's mind, to maintain the Pythagorean symmetries: Seven notes in the octave, seven Wanderers in the sky, seven colors in a ray of light, Cicero's cosmic seven that revealed the thoughts of God.

As mankind continues to explore interplanetary space, we keep stumbling upon harmonic intervals and geometric patterns. Neptune and Pluto exhibit a 2:3 orbital resonance, while the Lagrange points between celestial bodies adhere to precise geometric alignments.

The Pythagoreans would be delighted.

Celestial Twins

AT THE GATES OF HEAVEN

COIN NEWS numismatic monthly, June 2012

IS THERE A COIN OF THE ROMAN Republic more artfully executed than the denarius of C. Servilius (136 BC) that depicts the divine Twins astride prancing horses, while the two bright stars of the constellation Gemini dance above their heads? Why lavish so much talent and attention on a representation of the Dioscuri that is destined to wear down so soon? And why do the spears of the Twins cross in that peculiar manner behind them, by the double stars?

From its inception, the denarius of the Republic bore the image of the Dioscuri who, according to legend, saved the fledgling city from destruction at the battle of Lake Regillus and appeared miraculously in the Forum at the fountain of Juturna to announce the glad tidings. For centuries, the silver coins of the Republic paid grateful homage to the savior Twins, patrons of the equestrian military order that stood between the golden senators and the brassy plebs in the streets.

The celestial duo were guardians of the celestial portals because their twin stars, in the constellation Gemini, stand by the intersection of the Milky Way and the Zodiac. The Roman author Macrobius reveals that the intersections of these two celestial circles indicate the very Gates of Heaven, a neo-Platonist tradition that stretches back from Macrobius to Cicero, and thence to Plato's cosmic X in his Timaeus (see Coin News, January 2012: 'Plato's X on Roman Coins').

In a playful reference to that location in the heavens, a denarius of C. Antestius (146 BC) has a small dog running below the Dioscuri riding horses. On a celestial map, that diminutive canine would be the constellation Canis Minor that runs along the heavens right below Gemini. Who wouldn't be delighted by such clever imagery?

Denarius of Antestius with the Dioscuri (the constellation Gemini) accompanied by a small dog (the constellation Canis Minor).

The Romans had inherited the cult of the savior duo from the Greeks, who placed the Dioskouroi on coins from Magna Graecia to Bactria, stretching over 3000 miles. Halfway in between, on coins of Chalkis, Syria, the twins stand side-by-side and each extends a leg to form an X, alluding to the celestial intersection by the feet of the Gemini in the heavens.

44

How far back did this tradition extend? One of the earliest Greek poems put to paper, the 'Homeric Hymn to the Dioskouroi,' already invokes the savior Twins around 600 BC:

Quick-glancing Muses, sing of Zeus' sons,
The Tyndaridai, splendid children of fair-ankled Leda,
Horse-taming Kastor and blameless Polydeukes.
She mingled in love with Kronyon, lord of dark clouds,
Under the peak of Taygeros, that lofty mountain,
And bore these children as saviors of men on this earth...

The Dioskouroi, the twin 'sons of god,' were the children of the Spartan queen Leda, who received the amorous attentions of the king of the gods. Zeus descended to the mortal plane in the form of a swan, the constellation Cygnus. His sons, the divine twins, having finished their exploits on earth, ascended to the heavens as the constellation Gemini, earning the right to guard the celestial gates.

Although the celestial duo had received due reverence in the Greek world and the Roman Republic, the coming of the Empire tarnished the luster the Gemini once enjoyed. The hubris of the emperor of Rome knew no bounds and, as Pontifex Maximus, high priest of the Roman state cult, he appropriated for himself control of the heavenly portals. This scheme would work for three centuries, but when the Empire began to crumble, the celestial Twins would again be implored, and their image and symbols (the pilei, or caps, and the twin stars) again appeared on Roman coins.

In the days of the Tetrarchy, the co-emperor Maxentius gained control of the city of Rome, but not much else of the Empire. To legitimize his claim to imperial authority, Maxentius minted coins that showed the Dioscuri holding their horses while the twin stars of Gemini shine above their heads, advertising Rome's traditional control of the heavenly gates. More advanced and tantalizing symbolism appears on a variation of this coin where we see the Dioscuri in the same setting, but with the legendary she-wolf at their feet, suckling the abandoned twins who long ago founded the Eternal City.

45

Here Maxentius directly links Romulus and Remus to the heavenly Twins, underscoring Rome's stewardship of the celestial portals, and proclaiming that the divine gates are in his hands.

Coin of Maxentius with the Dioscuri standing guard over the Lupa Romana suckling the twins Romulus and Remus.

Most likely, it was this unwise self-aggrandizement that goaded Maxentius' co-emperor, Constantine, to march upon Rome in AD 312. Against the battle-hardened legions of Constantine, the troops of Maxentius stood little chance of success. Relentlessly pushed back at the Battle of the Milvian Bridge, they fled in panic and many drowned in the Tiber, as did their commander Maxentius.

Recently, as I was telling this long-winded story to my patient wife, she cut me short: "So this was the Battle for the Gates of Heaven!" Stunned, I had to admit that she had put it altogether in a nutshell. By capturing Rome and its heavenly portals, Constantine secured his hegemony in imperial politics and eventually steamrolled all others on his march to be the one and only Emperor.

Poor Maxentius did not survive the internecine feeding frenzy of the co-emperors, but the sanguine Constantine and his heirs learned a valuable propaganda lesson from him. On commemorative coins of the period, the legend VRBS ROMA encircles the bust of Roma, while the reverse shows the Lupa Romana suckling the twins below two stars, specifically linking the twins who founded Rome to the twin stars that stand by the celestial portals, Castor and Pollux.

Though Constantine would build a new imperial capital

by the Bosporus, Rome retained its heavenly connection, and when the Christians began to flex their political muscle, they placed the Chi Rho sign between the twin stars of the Dioscuri on VRBS ROMA coins, thus appropriating for themselves the centuries-old symbol of the heavenly gates, whose control was now in Christian hands.

Commemorative Urbs Roma coin with the Lupa Romana suckling the Twins that founded Rome, while the twin stars of Gemini shine above and the Chi Rho sign sits between them.

The ancient astral beliefs achieved one last flare-up under Julian the Philosopher (emperor AD 361-363), who sought to revive the rites and rebuild the temples that had been destroyed by Christian zealots. He had studied with the philosophers of Athens, and he sought initiation into what remained of the Mysteries of Mithras and of Eleusis. The enigmatic coin that best captures Julian's ancestral traditions shows a bull standing, with twin stars shining above.

Coin of Julian the Philosopher, showing a bull (Taurus) with twin stars above (Gemini).

Generally, this has been explained as representing an Apis bull, but the two stars on high open a window to a simple astronomical interpretation: the twin stars stand for the constellation Gemini and the bull represents its neighbor along the ecliptic that defines the Zodiac, the constellation Taurus.

Why do these two particular constellations appear on the coin of Julian? Because the Milky Way intersects the Zodiac right between Gemini and Taurus: at this intersection stand the Gates of Heaven guarded by the divine twins, the Dioscuri, the Gemini. Julian's coin in fact points to the specific location of the Celestial Portals, as do all ancient coins sporting the Dioscuri, or their twin caps, or their dual stars.

How far back does the tradition of twin guardians at the heavenly portals go, beyond Rome and archaic Greece? As far back as ancient Babylon, where the myth of Adapa (c. 1300 BC) guides its protagonist to the celestial gates guarded by two gods. One of these guardians is named Dumuzi or Tammuz, which means 'twin,' proving that the stories of a celestial duo at the heavenly gates is one of the oldest astronomical myths of mankind, and would be one of its favorite themes on ancient coins for centuries.

Demeter's Torch & The Mysteries of Eleusis

THE CELATOR numismatic monthly, March 2012

ON ANCIENT GREEK COINS, THE EAR of grain that sustained mankind (our daily bread) often stood for the goddess Demeter, and she also brought spiritual sustenance to mortals when the Mysteries were established at Eleusis, near Athens. The Homeric *Hymn to Demeter* recounts the founding myth of the Mysteries, and at the beginning we find Demeter carrying torches as she searches for her daughter Persephone, who has been abducted by Hades (Death), the brother of Zeus.

> Then for nine days divine Demeter roamed over the earth,
> holding torches ablaze in her hands…

At Eleusis, the torches that Demeter carried would become the symbol of humanity's search for what lies beyond. All who spoke Greek were welcome to the Mysteries – women, men, even slaves – and later all citizens of the Roman Empire could receive initiation, as many Roman emperors did.

The Greek colonies of Magna Graecia displayed a special fondness for the Eleusinian rites. In Sicily, Syracuse erected a temple to the Two Goddesses (mother and daughter), while Menaion issued coins showing Demeter and crossed torches, a motif that would also be found at Eleusis (Figure 1).

Figure 1. Left: Coin of Menaion, Sicily (2nd cent. BC). Obverse: Veiled head of Demeter. Reverse: Crossed torches. Right: Crossed torches at Eleusis.

Up on the Italian peninsula, the coins of Metapontum presented a unique version of the Eleusinian torch: intersecting branches atop a central staff, a figure now called the 'cross torch' (Figure 2).

Figure 2. Coins of Metapontum with the ear of grain of Demeter. Top: helmeted head of Leukippos, founder of Metapontum, cross torch at left (c. 300 BC). Bottom: Ivy-crowned head of Dionysos; on reverse, cross torch at right (c. 200 BC).

The cross torch of Eleusis is also found on red-figure vases of south Italy that depict the abduction of Persephone, with the mother desperately searching for her daughter (Figure 3), dramatizing the pivotal crisis of the Eleusinian myth.

Figure 3. Demeter brandishes the Eleusinian cross torch as she searches for her daughter (Apulian hydria, c. 330 BC; NY Metropolitan Museum of Art).

The Mysteries of Eleusis also played out the connection between Demeter (bread) and Dionysos (wine), as seen above on a coin of Metapontum (Figure 2, bottom), and as expressed in Sophocles' *Antigone*, written around 442 BC:

Thou who dost rule over Italia's pride
and at Eleusis in Deo's bosom wide
dwellest, Deo, the Mother of all,
Bacchos, Bacchos, on thee we call.

The Lesser Mysteries were held at Agra (purportedly commemorating the death and resurrection of Dionysos) while the Greater Mysteries of Demeter (celebrating the return of Persephone and the rebirth of Nature each year) took place at Eleusis, so that supplicants would have a chance to attend the lesser rites before rising to the higher level.

The Romans adopted much of the Greek culture of Magna Graecia, and the Mysteries would be captured on coins of the Roman Republic where on one side we find Liber/Bacchus (Dionysus), while on the reverse Ceres (Demeter) holds the torches that signify her cult at Eleusis (Figure 4).

Figure 4. The Eleusinian Mysteries on denarii of the Roman Republic. Top, obverse: Liber (Bacchus) wreathed in ivy (Lesser Mysteries); reverse: Ceres in chariot drawn by serpents, carrying torches (Greater Mysteries). Bottom, obverse: Bacchus (vine); reverse: Ceres with torches, behind a plough (grain).

The great jurist and statesman Cicero (who had been initiated at Eleusis) best characterized the Roman attitude toward the Mysteries:

> Among the many excellent and divine institutions that your Athens has developed and contributed to human life, there is none, in my opinion, better than these mysteries, by which we have been brought forth from our rustic and savage mode of existence, cultivated and refined to a state of civilization; and as these rites are called "initiations" so, in truth, we have learned from them the first principles of life and have gained the understanding, not only to live happily, but also to die with better hope. – Cicero

For that death with a "better hope" during the Roman Empire, Augustus was initiated at Eleusis in 31 BC, as were many later emperors who often put Ceres on their coins (Figure 5).

Figure 5. Roman Imperial coins with Ceres on reverse, holding symbols of the Mysteries of Eleusis. Top: Dupondius of Claudius (c. 50 AD). Middle: Denarius of Domitian (c. 85 AD). Bottom: Sestertius of Hadrian (c. 130 AD).

The empress Faustina received a wreath at Eleusis, as did her adopted son Marcus Aurelius, who would write the inspirational *Meditations* and who rebuilt the Temple of Demeter after it was sacked in 170 AD. Faustina was a fervent devotee of the Eleusinian Mysteries, with their promise of a blessed afterlife. When she passed away, Antoninus Pius had his wife deified (DIVA) and issued coins that testified to her hopes and beliefs (Figure 6).

Figure 6. Denarii with obverse of Diva Faustina. Top, reverse: Ceres with ears of grain and long torch, symbols of the Mysteries of Eleusis. Bottom, reverse: CONSECRATIO issue with Ceres holding torch.

Here we find the deified Faustina greeted by Ceres holding ears of grain and a torch (symbols of the Eleusinian Mysteries), and we also see her welcomed to the divine abode by Ceres/Demeter holding a torch, with the legend CONSECRATIO above.

With this sequence, we witness the deep imprint that the Mysteries of Eleusis left upon Western civilization: the promise of a beatific afterlife became a central tenet of Christianity, and Eleusinian symbolism would be invoked in the Gospels.

> Unless a kernel of wheat falls to the ground and dies, it remains only a single seed. But if it dies, it produces many seeds. – John 12: 24

> I am the true vine. – John 15:1

With this ancient connection readily apparent, it would be difficult to admire Leonardo da Vinci's *Last Supper*, or to attend Mass in a church, and not be reminded of the Mysteries of Eleusis.

Plato's X on Roman Coins

COIN NEWS numismatic monthly, January 2012

EVER COME ACROSS A COIN WHOSE enigmatic symbolism leaves you mystified? Many Roman coins, for example, show globes or spheres that proclaim Rome's dominion over the universe, but their telling features have often been eroded past recognition.

So when the celestial sphere on a Roman coin bears clearly visible intersecting lines, a chill might legitimately run up your spine. And when these intersecting lines appear on coins minted over hundreds of years, it might make your hair stand on end. What could the intersecting lines on a cosmic orb possibly stand for? Fasten your seatbelts – could be a bumpy ride.

One of the great paintings in the Vatican, Rafael's 'School of Athens,' portrays Plato holding his book *Timaeus*, a most

Pythagorean work that encompasses the Cosmos through mathematic proportions, musical intervals and geometric figures. Since Pythagoreans ostracized those who revealed their secrets to the uninitiated masses, much of Plato's cosmology can only be glimpsed, as he'd say, "through a glass dimly."

In *Timaeus*, Plato describes the World Soul as having the shape of an X (the Greek letter chi) formed by the intersection of two celestial circles. One of these circles, Plato tells us, is the course of the Wanderers, the path of the Planets along the ecliptic that defines the constellations of the Zodiac. The other circle, part of the celestial lights that do not change, our reticent author does not name. But there is a golden chain that, if followed carefully, can lead us to the mysteries that Plato guarded so carefully.

In the twilight days of the Empire, the Roman author Macrobius wrote his *Commentary on Cicero's Dream of Scipio*, a neo-Platonist exposition of Hellenistic philosophy, where he explains that the intersections of the Zodiac (the path of the Planets) and the Milky Way indicate the very Gates of Heaven.

> The soul descends from the place where the zodiac and the Milky Way intersect... Souls are believed to pass through these portals when going from the sky to the earth and returning from the earth to the sky... – Macrobius, *Commentary on Cicero's Dream of Scipio*

Macrobius paints a cosmos where the Milky Way is the celestial abode of souls, spiritual entities that descend and return along the path of the Planets, the ecliptic that marks out the Zodiac and intersects the Milky Way, forming the letter X (*chi*) in the sky. At the intersection of these two circles stand the celestial portals.

Where might Macrobius have gleaned this jewel of ancient soteriology? From 'The Dream of Scipio,' the text he is commenting on, where the Roman statesman Cicero makes the Milky Way the home of virtuous souls and invokes the Pythagorean harmony of the spheres, the music that the Planets produce as they travel along the Zodiac. In effect, Cicero presents his reader (like Macrobius) with two intersecting celestial circles.

Where might Cicero have found this notion about the celestial intersections? In the works of the master himself, Plato, whose Pythagorean texts Cicero translates (*Timaeus* with its cosmic X) and emulates (the 'Myth of Er' at the end of *Republic*). This cosmology was not unique to Cicero and his neo-Pythagorean mentor, Nigidius Figulus, for the Via Galactica as the abode of sages and heroes also appears in Manilius' *Astronomica*, an astrological handbook from around the time of Augustus, and there too we find the intersections of the Planets along the Zodiac with the Milky Way, where Manilius places Plato in the heavens.

Confidently then, we can trace this tradition back from Macrobius (*Commentary on The Dream of Scipio*) to Cicero (whose 'Dream of Scipio' at the end of his *On The Republic* was a reworking of the 'Myth of Er' at the end of Plato's *Republic*) and thence to Plato himself, a tradition that reached back at least seven hundred years.

How can we positively know that Plato's X in *Timaeus* was common knowledge throughout the Roman Empire? From the testimony of the early Christian apologist Justin Martyr (d. AD 165) who argued that Plato's X in the heavens was, in fact, a foreshadowing of the Christian cross.

> And the physiological discussion concerning the Son of God in the Timaeus of Plato, where he says, "He placed him crosswise in the universe…" – Justin Martyr, *First Apology*

Justin obviously expects his readers, educated citizens of the Roman Empire, to be familiar with Plato, with his work Timaeus, and with the celestial X described therein, proving that Plato's cosmic intersection, the Chi in the sky, was well-known throughout the Hellenistic world.

Giving evidence that Plato's X appeared on coins of the Roman emperors, Justin writes of the celestial intersecting symbol:

> With this form you consecrate your emperors, and declare them gods with inscriptions.

The intersecting shape that Justin invokes can be seen on Roman *consecratio* coins where the departed emperor is declared divine (*divus*), while the bird of Jupiter sits atop a celestial sphere with intersecting lines, declaring Rome's control over the Gates of Heaven.

Left: Brass coin of Titus, with Divus Augustus on obverse, and imperial eagle atop celestial orb with intersecting lines. Right: *Consecratio* denarius acclaiming Divus Marcus Aurelius; on reverse, the Roman eagle sits on the heavenly sphere with celestial crossroads, Plato's X.

The emperor was commander-in-chief of the Roman army, but more importantly he was Pontifex Maximus, the highest priest who controlled access to the heavenly realm, much as the current pontiff in Rome holds the keys to the celestial kingdom.

From where did such ideas possibly originate? Not surprisingly, the heavenly portals with gatekeepers can already be found in Plato's 'Myth of Er,' from around 300 BC:

> There were two openings in the earth, and above them two others in the heavens, and between them judges sat. These, having rendered their judgment, ordered the just to go upwards into the heavens through the door on the right...– Plato, *Republic*

Plato's most Pythagorean texts, the 'Myth of Er' and Timaeus, refer to the same thing: to the heavenly gates whose location Macrobius would divulge centuries later and which, as Plato's X, would appear on coins of Roman emperors, bolstering their claims of near-divinity and proclaiming their god-given right to rule over the entire Cosmos.

Titus minted coins with Divus Augustus on one side and the imperial eagle atop the celestial sphere inscribed with

Plato's X on the reverse. After the death of Titus, Domitian deified his own son who had died in infancy, placing him atop the heavenly orb by the celestial crossroads, while seven stars (the Planets along the ecliptic) danced overhead as playthings of the divine child.

Antoninus Pius enthroned Italia triumphant on the celestial sphere dotted with stars and sporting intersecting lines, while his heir Marcus Aurelius would write the stoic Meditations and mint coins with Plato's X on the cosmic orb.

CELESTIAL INTERSECTIONS ON ROMAN IMPERIAL COINS. Top, left. Denarius of Domitian shows his deified son atop a celestial sphere with intersecting lines and playing with seven stars, the Planets. **Top, right.** Sestertius of Antoninus xPius, with Italia sitting on a heavenly orb with intersecting lines and visible stars. **Middle, left.** Denarius of Antoninus Pius, with Italia sitting on celestial globe with intersecting lines. **Middle, right.** Denarius of Marcus Aurelius, with Providentia pointing to a celestial orb at her feet that shows intersecting lines. **Bottom, left.** Coin of Constantine with Jupiter presenting a heavenly globe with intersecting lines and dotted with stars. **Bottom, right.** Coin of Gratian (c. 383 AD) showing Roma enthroned, holding a celestial sphere with intersecting lines, the gates of heaven.

Later emperors would follow these traditions, but none as enthusiastically as the house of Constantine. On coins of Licinius and his co-emperor Constantine, we witness Jove proffering the celestial globe with intersecting lines and dotted with stars. We see this same symbol when Sol Invictus offers the cosmic sphere to the emperor, with Plato's X clearly visible. And the next generation, the sons of Constantine, would carry this symbolism further, perching the mythic Phoenix atop the heavenly sphere engraved with Plato's intersection and dotted with stars.

Even the Christian emperor Gratian (c. AD 383) placed Roma enthroned on his coins, holding the celestial orb with an X and dotted with stars, signaling Rome's control of the heavenly portals. Gradually Plato's Chi (X), representing the gates of heaven, would morph into the Christian Chi Rho...

But that, as they say, is another story.

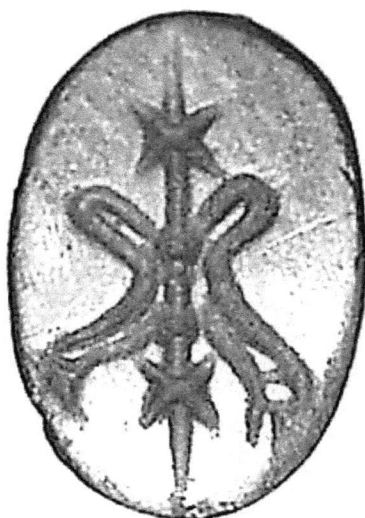

Dionysos

The Mysteries Made Visible

Parabola quarterly journal, Fall 2011

A SHORT WHILE AGO, MY twelve-year-old grandniece asked a very astute question at the dining table: "How come we can see Jupiter and Venus, but Dionysos is not visible, like other ancient gods?"

That question had been rattling around in my mind for quite some time and her query ignited a surprising remembering, which according to Plato is the process of learning. As it turns out, Dionysos *is* linked to a celestial phenomenon, but one so rarely seen that, in antiquity, it spawned the Mysteries and remains a mystery for most people today.

In his *Bacchae*, the Athenian playwright Euripides has Dionysos manifesting a visible light reaching from the earth to the sky, while in *Antigone*, Sophocles claims that the god leads the dance of the stars, indicating that Dionysos does have a connection to phenomena visible in the firmament.

> Dionysos cried out: Maidens,
> I bring the man who makes a mockery of you
> And me and my orgies; take vengeance on him!
> As he addressed them, a light of awesome fire
> Was fastened on the heaven and the earth.
> – Euripides, *Bacchae*
>
> God of many names…
> O Bacchus…
> O leader in the dance of the stars
> That circle in the night…
> Come, O Lord…
> – Sophocles, *Antigone*

Now, even amateur astronomers know that the stars do not dance. They are stationary, fixed in their position relative to each other due to their great distance from us. The 'stars' that do dance are the Wanderers of our planetary system, the seven visible bodies that course along the Ecliptic and mark out the constellations of the Zodiac. For Dionysos to lead the dance of the Wanderers, he too would have to travel along the Ecliptic and soar along the Zodiac.

Thank heavens, there *is* a celestial phenomenon that fits what Euripides and Sophocles revealed, an awesome light in the night sky that follows the Ecliptic: the Zodiacal Light. Composed of myriads of dust particles that circle the Sun along the path of the planets and that, like the planets, reflect the solar glare, this magic dust becomes visible at specific times of the year and at particular hours of the night, revealing to the initiated a golden pathway to the heavens. And when planets are propitiously located along this visible path, we witness the celestial stairway leading up to the divine abode that formed the basis of ancient astral religion (Figure 1).

Figure 1. Planetary stairway along the zodiacal light: Bright Venus at the bottom, Mars in the middle, Saturn at the top (Photo: Tunc Tezel/TWAN).

The Greek poet Pindar painted the glowing heavenly staircase in his *Odes* (c. 400 BC), while the description of this phenomenon hundreds of years earlier by the greatest Greek poet would become famous as the Golden Chain of Homer.

> First did the Fates in their golden chariot bring heavenly Themis, wise in counsel, from the springs of Ocean to the awesome stair that marks the shining way to Olympus... – Pindar, *Odes*

> Hangs me a golden chain from heaven, and lay hold of it all of you, gods and goddesses together – tug as you will, you will not drag Zeus the supreme counselor from heaven to earth. – Homer, *Iliad*

In the *Ten Books On Architecture* of the Roman writer Vitruvius (which inspired Leonardo's Vitruvian Man), we also find the planetary staircase, while the Greek philosopher Celsus gave his understanding of the Mysteries of Mithras, a cult that attracted many adherents from the Roman army and promised the ascent of the soul through a planetary ladder (Figure 2).

> The Moon, Mercury, Venus, the Sun, as well as Mars, Jupiter, and Saturn, differing from one another in the magnitude of their orbits as though their courses were at different points in a flight of steps... – Vitruvius

> In that system there is an orbit for the fixed stars, another for the planets and a diagram for the passage of the soul through the latter. They picture this as a ladder with seven gates, and at the very top an eighth gate... – Celsus, *On the True Doctrine*

Figure 2. Mosaic path at Ostia mithraeum, showing the planetary ladder to heaven (Photo: Payam Nabarz, *The Mysteries of Mithras*).

The symbol of the planetary ladder to the heavens was carried at the forefront of Roman legions on standards that bore circles, disks and crescents representing heavenly bodies (Figure 3, top). The first Roman emperor Augustus adopted the zodiacal sign Capricorn as his personal symbol and, after he was deified, this sign would appear atop legion standards hundreds of years later (Figure 3, bottom). The promise of a heavenly ascension inspired legionnaires as they marched into battle, with Divus Augustus (Capricorn) ready to welcome them into the celestial abode.

Figure 3. Top: Denarius of Marc Antony, showing standards with Moon crescent at the bottom, topped by circles or orbs. Bottom: bronze coin of Gordian III; on the reverse Capricorn, the astrological sign of Augustus, sits atop the flanking standards.

Monotheistic religions tried to stamp out all evidence of ancient planetary cults, but traces survived. The Great Menorah of the Temple of Jerusalem that was carried off by Roman soldiers (as seen on the Arch of Titus in Rome) had seven branches, and the Jewish writers Josephus and Philo of Alexandria assert that these seven lights stood for the Planets. The consecutive lighting of menorah candles over days at Hanukkah replays the ascent through the Planets, with the first candle to be lit named Shamash, the ancient mid-Eastern Sun god.

Mirroring this symbolism, many Christian churches today have seven lamps before the altar, from the *Revelation* of John of Patmos, seven lights that represent the Planets. And when Muslim pilgrims arrive in Mecca, their first duty is to circumambulate the Ka'aba: Briskly for the first three circuits (the inner Planets: Moon, Mercury, Venus), and then more leisurely for the remaining four circuits (the outer Planets: Sun, Mars, Jupiter, Saturn), paralleling the ancient ascent through the Planets.

In *Paradiso*, the 14th-century Italian poet Dante describes a climb through the Planets, proving that this tradition lived on for many generations. But the most amazing survival of the planetary ascent to the heavens is our Week, the seven-day voyage through the Planets. The Roman historian Dio Cassius reveals that the order of the days of the week is based on musical intervals, and so if we follow the classical planetary sequence of Vitruvius, we start with the Moon (Monday, the first day of the week), skip 3 planets (the musical interval of the Fifth) and arrive at Mars (French: Mardi). Repeating the procedure, we circle back to Mercury (French: Mercredi), then on to Jupiter/Jove (French: Jeudi), to Venus (French: Vendredi), then to Saturn (Saturday), and finally arrive at Sunday. Then we repeat the uphill climb through the Planets all over again, much like Sisyphus.

The Roman encyclopedist Pliny ascribed the Music of the Spheres to Pythagoras, while hundreds of years later the jurist Martianus Cappela described an allegorical harmonic ascent through the planetary spheres.

> Occasionally Pythagoras draws on the theory of music, and designates the distance between the earth and the moon as a whole tone, that between the moon and Mercury as a semitone, between Mercury and Venus the same, between her and the sun a tone and a half, between the sun and Mars a tone, between Mars and Jupiter half a tone, between Jupiter and Saturn half a tone… – Pliny, *Natural History*

> Philology ascended rapidly from here and flew by a half tone as far as the circle of Venus…Soon she was eager to make the toilsome journey to the sun's circle – an ascent rendered toilsome by its distance of three half tones, or a tone and a half. – Martianus Capella, *Marriage of Philology and Mercury*

In his *On The Republic*, the Roman orator Cicero mimics Plato's *Republic* and invokes a return to the celestial regions, a return because the soul supposedly descends from the heavens acquiring different attributes through the planetary spheres (according to Macrobius), attributes it will relinquish in reverse order once free of the material body.

> What is this great and pleasing sound that fills my ears? "That," replied my grandfather, "is a concord of tones separated by unequal but nevertheless carefully proportioned intervals, caused by the rapid motion of the spheres themselves... Gifted men, imitating this harmony on stringed instruments and in singing, have gained for themselves a return to this region, as have those who have devoted their exceptional abilities to a search for divine truths." – Cicero, *Dream of Scipio*

> In the sphere of Saturn it [the soul] obtains reason and understanding... in Jupiter's sphere, the power to act... in Mars' sphere, a bold spirit... in the sun's sphere, sense perception and imagination... in Venus' sphere, the impulse of passion... in Mercury's sphere, the impulse to speak and interpret... and in the lunar sphere, the function of molding and increasing bodies. – Macrobius, *Commentary on the Dream of Scipio*

The return of the soul to the celestial regions along the planetary path was the great secret of ancient mystery religions, whether the Mysteries of Mithras or the Mysteries of Eleusis, and since the Planets must travel beneath the earth, this journey inevitably encounters Death (Hades).

Dionysos descended to the underworld and returned, just as Herakles came back with three-headed Kerberos, and for this they were called *soter*, or *savior*, as guarantors of the soul's safety in the afterlife (Figure 4, top). Just skirting sacrilege, Aristophanes comically re-invented Dionysos' otherworldly journey in *Frogs*, where Dionysos asks Herakles for advice on how to reach Hades. Prior to his voyage to the underworld, Dionysos was initiated into the Mysteries of Eleusis (as was Herakles), and the connection between the Eleusinian goddess and Dionysos would survive for centuries.

> God of many names,
> Glorious child of Thebes,
> Whose mother was bride
> To Zeus' deep thunder!

It is you who guard the fame of Italy,
You who look after the embrace, at Eleusis,
Of Demeter, all-welcoming goddess.
O Bacchos...
    – Sophocles, *Antigone*

When the Greeks colonized southern Italy (Magna Graecia), the city of Metapontum minted coins that show Dionysos on one side, while the reverse has the ear of grain of Demeter flanked by the four-headed torch of Eleusis (Figure 4, middle). Similarly, silver coins of the Roman Republic show Liber (Dionysos) on one side, while Ceres (Demeter) carries torches on the reverse as she searches for her abducted daughter (Figure 4, bottom), a pairing that cements the mystical relationship between Demeter (grain, bread) and Dionysos (grape, wine).

Figure 4. Top, coin of Thrace: Ivy-crowned head of Dionysos, on reverse Dionysos Soter (*Savior*).

Middle, coin of Metapontum: Head of Dionysos, on reverse the ear of grain of Demeter flanked by the four-headed cross torch of Eleusis.

Bottom, silver denarius of the Roman Republic: Head of Liber (Dionysos), on reverse Demeter searches for her daughter, brandishing torches in a chariot drawn by serpents.

The influence of the Mysteries of Eleusis on the Roman world was so pervasive that Octavian asked to be initiated there after his victory over Marc Antony at Actium, acquiring the right to be called Augustus and to be deified after his death. Later Roman emperors and empresses, in order to join the visible gods in the heavens, would often make the pilgrimage to the village outside Athens where the mystic rites were held.

In his *Eleusis*, Carl Kerenyi amply documents Dionysos' link to the Eleusinian Mysteries, rites that were celebrated in the Spring (Lesser Mysteries) and in the Fall (Greater Mysteries), rites celebrated at night. It is precisely at these times of the year that the zodiacal light is best seen from temperate latitudes, either after sunset, or before sunrise.

Pindar had pointed to the visible manifestation of Dionysos at harvest time, while dithyrambs and cultic hymns implored the god to appear, often in a stellar context:

> May Dionysos, bringer of joy, foster the grove of trees,
> The holy light at summer's end.

> Come, O Dithyrambos, Bacchos, come,
> Euios, Thyrsos-Lord, Braites, come...
> Whom in sacred Thebes the mother fair,
> She Thyone, once to Zeus did bear.
> The stars danced for joy...

How could the god possibly appear when the stars are dancing? By manifesting himself as a holy light across the night sky that only those initiated into the Mysteries would recognize and understand.

In *Phaedo*, Plato alludes to the esoteric content of these rites: "Many carry the thyrsus, but few are Bacchi." The dance of the followers of Dionysos, like the dance of the Whirling Dervishes, is the dance of the Planets along the Ecliptic, along the mystic light that appears only at specific times of the year.

In order to witness this ethereal light for myself, I flew down to Arizona where the desert night skies are still dark at some distance from light polluting cities. An hour after sundown, as I drove along pitch-black roads searching for my bed-and-breakfast, I realized that this was the witching hour. Pulling over and turning off all lights, I waited for my eyes to adjust to the darkness.

And there, ever so faint but undeniably visible, arched on high the Zodiacal Light, the celestial pathway along which Dionysos leads the dance of the whirling Planets.

Celestial Symbols on Roman Standards

THE CELATOR numismatic monthly, June 2011

ON ROMAN COINS, WE OFTEN SEE the standards that were carried in front of the legions by the bravest, strongest soldiers, dressed in the pelts of the fiercest animals, like Hercules in his lion skin. In battle, the standards communicated to the troops the commands of their leaders, and the greatest shame befell any legion that lost its standards.

75

One type of Roman standard had several disks, or orbs, stacked one atop the other along a central pole. What did these disks represent? In *A Dictionary of Ancient Roman Coins*, John Melville Jones writes:

> Standards were decorated with a variety of objects on their shafts. Paterae, phalerae, crescents and circles (perhaps lunar and solar symbols)…

If the crescent logically stands for the Moon, then the orbs might represent other celestial objects that move along the ecliptic: the Planets, with two of the Wanderers being the Moon and the Sun (explaining the "lunar and solar symbols"). Already on coins of Mark Antony (c. 30 BC), we see standards with the lunar crescent at the bottom and several disks, circles, or orbs above them (Figure 1).

Figure 1. Denarius of Marc Antony, showing standards with Moon crescent below stacked orbs.

By the early Roman era, major gods and goddesses were linked to the wandering Planets and their heavenly orbits, with coins of the Republic depicting Venus (Figure 2, top) as well as Luna riding in their celestial bigas, while Sol, Mars (Figure 2, bottom), Jupiter, and Saturn galloped overhead in their quadrigas.

**Figure 2. Denarii of the Republic. Top: Venus in her heavenly biga (c. 133 BC).
Bottom: Mars in his celestial quadriga (c. 131 BC).**

Opinions differed as to the order of the Planets (the military cult of Mithras, for example), so we do see coins that have the lunar crescent at other locations along the standards (in multiple, at times). But the most prevalent view held that since the Moon is one of the largest Wanderers in the sky and occasionally eclipses the Sun, it must be the closest to the Earth, thus lowest on the pole. Plato already presents this lunar location in *Timaeus*:

> He placed them into the orbits traced by the period of the Different...
> He set the Moon in the first circle around the earth...

The days of the week were named after the planetary gods (Saturn day, Sun day, etc.), and fittingly the first day of the week is Moon day (French: lundi). Our entire week is a voyage through the Planets (along musical intervals according to Cassius Dio), with the neo-Platonist writer Martianus Capella describing a harmonic ascent in *The Marriage of Philology and Mercury* (c. 400 AD), and here too the first step up is to the lunar sphere:

> Then the bearers picked up the goddess' palanquin and with great effort carried her aloft. Borne up by their buoyancy they rose 126,000

stades, and completed the first of the celestial tonal intervals; then the maiden entered the circle of the moon...

Why would Roman legions carry at their forefront a representation of the Planets stacked along the ecliptic? Because savants of the time, neo-Pythagoreans (Nigidius Figulus, Cicero, Numenius) as well as neo-Platonists (Porphyry, Iamblichus, Macrobius), taught that the departed soul ascended to the celestial abode through the planetary spheres. This elite soteriology would be passed down to the Roman legions by the scions of the equestrian and senatorial orders, the traditional leaders of the Roman army. Without doubt, the greatest glory and reward for a Roman warrior, should he die that day on the battlefield, would be to join the company of the heavenly gods.

Who would welcome this brave soldier upon his arrival in heaven? Following his *consecratio*, that would be the deified Octavian himself (Divus Augustus), the founder of the Empire ("pater patriae"), whose astrological sign Capricorn appears on his own coins as a personal *genius* (Michael Molnar, *Celator Magazine*, April 1994), just as it would appear at the top of Roman standards hundreds of years later (Figure 3). With the crescent Moon at the bottom and a zodiacal sign at the top, there can be little doubt that the orbs in between also stand for celestial objects that travel along the ecliptic: the Planets that trace out the Zodiac itself.

Figure 3. Coin of Gordian III (c. 240 AD). Reverse: Standards with Moon crescent at the bottom, planetary spheres in between, and the Zodiacal sign of Augustus, Capricorn (top left and right) flanking the imperial eagle, the bird of Jupiter.

In discussing celestial symbols on Roman standards, we must not forget the eagle that perches centrally with the greatest honors. As the funeral pyre of Augustus blazed, the bird of Jupiter was released to represent the emperor's soul flying to the heavens, a tradition that would be repeated for later emperors, as depicted on Roman coins where the deified emperor flies skyward atop an eagle (Figure 4).

Figure 4. Sestertius showing the soul of divus Marcus Aurelius carried to heaven on an eagle.

The notion of a heavenly ascent proved to be very popular in the Roman army, where the widespread cult of Mithras showed strong astrological affinities and portrayed a celestial ladder through the Planets (Figure 5, left) that would be described by the Greek philosopher Celsus:

> In that system there is an orbit for the fixed stars, another for the planets and a diagram for the passage of the soul through the latter. They picture this as a ladder with seven gates, and at the very top an eighth gate...

Away from light pollution, we can still witness the shifting planetary staircase that runs along the Ecliptic (the path of the Planets), and on auspicious nights the planets line up to reveal the stairway to heaven as the ancients saw it (Figure 5, right).

Figure 5. Left: Mosaic path at Ostia mithraeum depicting the ladder of the Planets (Photo: Payam Nabarz). Right: the shifting planetary staircase, with the bright planet Venus (bottom), Mars (middle), and Saturn (top) along the ecliptic with its zodiacal light (Photo: Tunc Tezel/TWAN).

The memory of the ascent through the Planets certainly survived for, in his *Paradiso,* the 14th-century Dante vividly recounts a passage through the planetary spheres. Earlier, around the time of Augustus, the Roman architect Vitruvius gives the classical order of the planetary staircase in his *Ten Books on Architecture*, while in the *Commentary on the Dream of Scipio* (c. 400 AD), Macrobius lists the attributes a soul acquires when descending to an earthly life, attributes that it will relinquish on its return journey through the planetary spheres when it leaves the body behind.

> The moon, Mercury, Venus, the sun, as well as Mars, Jupiter, and Saturn, differing from one another in the magnitude of their orbits as though their courses were at different points in a flight of steps... – Vitruvius

> In the sphere of Saturn it obtains reason and understanding... in Jupiter's sphere, the power to act... in Mars' sphere, a bold spirit... in the sun's sphere, sense-perception and imagination... in Venus'

sphere, the impulse of passion… in Mercury's sphere, the ability to speak and interpret… and in the lunar sphere, the function of molding and increasing bodies. – Macrobius

The treadmill of the Planets at the heart of astrology still supports many astrologers in our times that claim to reveal the future (remember Nancy Reagan's astrologer?), much as Roman emperors pored over the natal charts of potential challengers, and then had them executed to remove any possible threat.

Perhaps we have none other to blame for this than the greatest Western philosopher, Plato, who in *Republic* has the Fates harmonizing to the Music of the Planets.

And up above on each of the rims of the circles stood a Siren, who accompanied its revolution, uttering a single sound, one single note. And the concord of the eight notes   produced a single harmony. And there were three other beings sitting at equal distances from one another, each on a throne. These were the Fates, the daughters of Necessity… and they sang to the music of the Sirens. Lachesis sang of the past, Chlotho of the present, and Atropos of the future.

With the Fates singing of Time (especially of the future) to the music of the planetary Sirens, one gets a glimpse of the appeal of such a precise cosmology and soteriology, where the visible orbits of the Planets both reveal the future and erect a ladder to the celestial abode.

The planetary stairway to heaven on the standards of Roman legions delivered the ultimate physical challenge and spiritual promise, one that could well have issued from the mouth of a Roman commander: "Courage, men! We march into battle for fame, honor, and possibly, celestial immortality!"

The Cornucopia
& The Milky Way

The Celator numismatic monthly, accepted January 2013

THE FIGURE OF THE HORN OF PLENTY (*cornucopia*) graced
ancient coins from at least the Seleucid era to the Late Roman
Empire, promising abundance that rained down from above,
like manna from heaven (Figure 1).

Figure 1. Cornucopia on ancient coins. Top: Drachm of Demetrios I Soter (c. 150
BC). Bottom: Follis of Licinius, Constantine's co-emperor (c. 320 AD).

Where did this magic Horn come from? The Cornucopia originates from Greek theogony, the most basic level of myth that explains the origins of the cosmos as the birth and struggle for survival of the gods themselves. Hidden in a cave from his child-swallowing father Kronos, the infant Zeus was reared on the milk of a goat that belonged to the Muse Amalthea, or that represented the Muse herself (Figure 2).

Figure 2. Top, silver coin of Valerian II: The child Jupiter (Iovi Crescenti) rides atop the goat Amalthea and grasps her horn. Bottom, antoninianus of Gallienus: The goat Amalthea with legend Iovi Cons(ervatori) Aug(ustus).

When one of the goat's horns was somehow broken off, Zeus turned it into the Cornucopia, the horn that overflows with all things good and bounteous. And in gratitude for her lacteal services, Zeus placed the goat in the heavens as the bright star Capella, as reported by the Greek writer Aratus (c. 270 BC) and the Roman poet Ovid (c. 8 AD).

> If you wish to observe the Charioteer [Auriga],
> or have heard of Capella, the Goat...
> Inlaid on his left shoulder is the sacred Goat
> that according to legend nursed Zeus at her breast
> and is titled 'Olenian' by those who interpret
> Zeus and his heaven: a splendid bright star...
> – Aratus, *Phenomena*

On the first night of May
the star that tended Jupiter's cradle can be seen.
The rainy constellation of the Olenian She-Goat rises.
Heaven's her reward for the milk she gave…
He made stars of his nurse and her horn of plenty,
which even today is called Amalthaea's Horn.
    – Ovid, *Fasti*

Most revealing is the location of the bright star Capella in the night sky: it sits right by the Milky Way, the celestial milk with which the sacred Goat (Amalthea/Capella) nursed the baby Zeus, the future king of the heavens.

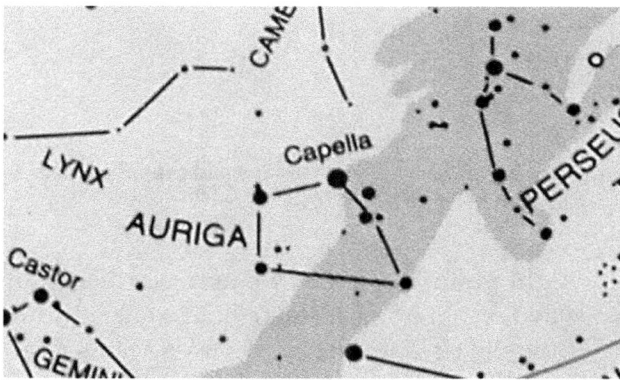

Figure 3. The bright star Capella (Goat) by the Milky Way.

In the logic of symbols, the equation runs as follows:
    Cornucopia = Amalthea + Goat + Milk = Capella + Milky Way.

The Via Galactica, the abode of gods and heroes (according to Cicero, Manilius, Martianus Capella, etc.), was the very source of divine largesse, and blessings poured down upon mortals through the enchanted Horn. Coins of Antoninus Pius give witness to the celestial nature of the Cornucopia when the figure of all-conquering Italia sits enthroned upon the cosmic sphere with visible stars, while cradling the Horn of Plenty (Figure 4).

85

Figure 4. Coins of Antoninus Pius (denarius and sestertius) with Italia on celestial globe holding the Cornucopia.

Due to light pollution, today we may not discern much of the Milky Way (except at dark sites), but in ancient times its awesome effulgence in the darkest sky was seen as the bounty that flows from the heavens, as the celestial milk that nursed the baby Zeus, and as the symbol of Zeus' gratitude and generosity, the Cornucopia.

GILGAMESH at the Gates of the Netherworld

FROM THE BULL OF HEAVEN TO THE SCORPION PEOPLE

IN THE BABYLONIAN EPIC, GILGAMESH and his adopted brother Enkidu defeat the giant Humbaba and kill the Bull of Heaven, but then Enkidu is laid low due to the ire of the scorned goddess Ishtar.

Gilgamesh proves inconsolable. He rages like a wild bull, attacking even lions in the desert. Eventually he sets out on a quest to understand mortality and the possibility of immortality. The first leg of the journey brings him to the mythical mountains where the Sun rises, to celestial gates guarded by Scorpion people.

> When he reached the mountain Mashu
> Which daily guards the coming out [of Shamash] –
> Their upper parts [touch (?)] the sky's foundation,
> Below, their breasts reach Arallu.
> They guard its gate, Scorpion-men
> Whose aura is frightful, and whose glance is death.[1]

Cuneiform tablets that tell the story of Gilgamesh surfaced during mid-19th century excavations of the Assyrian palace at Nineveh. Many versions of the myth, written in Sumerian and Akkadian, survive, as well as Hittite and Hurrian fragments and Sumerian prototypes.[2]

What made this story so special? Why was it passed on from generation to generation, from one culture to another, over thousands of years?

Astronomical Interpretation

One explanation for the longevity and popularity of the Gilgamesh myth is that it's all about the heavens, that the travels and travails all take place in the sky.

In the story, Gilgamesh worships the sun god Shamash, prays to the moon god Sin, while the goddess Ishtar, the planet Venus, would turn into a bitter enemy. Here we have the Wanderers in the sky determining the fate of mortals – the basis of astrology, a divinatory art first developed in Babylonia.

These Wanderers in the sky can be seen on *kudurrus* – land deeds witnessed by the gods and carved in stone – where we find along the top the starry symbol of Venus, the crescent of the Moon, and the radiant disc of the Sun (Figure 1).

Figure 1. Middle Babylonian kudurru (c. 1100 BC). At the top: Venus, Moon, and Sun, celestial Wanderers that travel along the ecliptic (British Museum).

The hypothesis that the Babylonian epic had an astronomical subtext is not new – de Santillana and von Dechend reached this conclusion in *Hamlet's Mill* in 1969: *"It becomes evident that all the adventures of Gilgamesh… are astronomically conceived…"*[3]

Since we've already encountered the celestial deities Shamash (Sun), Sin (Moon) and Ishtar (Venus), we can connect the Scorpion people that guard the heavenly gates to Scorpius, one of the constellations along the ecliptic, the course of the Wanderers as they delineate the Zodiac.

The location by the celestial scorpion has a special significance because it sits by the intersection of the Milky Way and the Zodiac (Figure 2). According to the Roman writer Macrobius, the Gates of Heaven are found at the intersections of these two celestial circles[4] – a tradition that can be traced over thousands of years through myths such as the Gilgamesh story in Babylonia, the tale of the Dioskouroi in Greece and Republican Rome, and the Mithras cult so popular in the Imperial army.

Figure 2. The center of the Milky Way intersects the Zodiac between Sagittarius and Scorpius: the Scorpion gate.

If Scorpius stands for one of the heavenly gates, the constellation that sits directly opposite on the Zodiac indicates the other celestial portal, the one by Taurus. Again, we find the intersection of the Milky Way and the Zodiac. The Bull of Heaven is the constellation Taurus and, on the other side of the Milky Way, Gilgamesh and his adopted brother (as the Gemini) face the celestial bull (Figure 3).

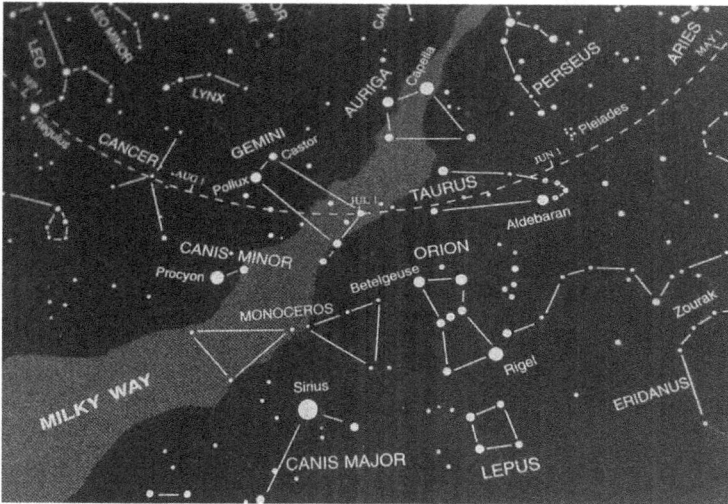

Figure 3. The Milky Way intersects the Zodiac between Gemini (the twin brothers) and Taurus (the Bull of Heaven): the Taurus gate.

Standing by the heavenly portals as savior gods in Greek and Roman religion, the celestial twins Castor and Pollux replay the Babylonian tale of Gilgamesh and Enkidu, where one of the brothers is mortal and dies, while the other one survives because he is a demigod.

In the same vein, the killing of the Bull of Heaven will become the cosmic sacrament of the Roman cult of Mithras – the tauroctony – showing the astronomical chain that links myths and cultures over thousands of years.

Celestial Gates and the Night Journey

The astronomical arc of the adventures of Gilgamesh starts with the gods that travel along the ecliptic (Sun, Moon, Venus), proceeds to the celestial intersection between the Brothers (Gemini) and the Bull of Heaven (Taurus), and arrives at the opposite heavenly gate guarded by Scorpion people (Scorpius).

The opposite locations of the portals in the sky are beautifully captured on a neo-Babylonian cylinder seal now in the Metropolitan Museum of Art in New York (Figure 4). Gilgamesh and Enkidu overcome the giant Humbaba, who is down on one knee. On the left, Gilgamesh sports the legs and tail of a bull-man (Taurus), while on the right of Humbaba, we see Enkidu with the tail and talons of a scorpion-man (Scorpius).

Figure 4. Gilgamesh (left, as a bull man), the giant Humbaba on one knee, Enkidu as a scorpion man. At right, a fish-clad priest reads the signs in the heavens (c. 700 BC, Metropolitan Museum of Art, New York).

The visual evidence here is clear to those familiar with the story and with the constellations visible at night: the tale of Gilgamesh and Enkidu encodes the two intersections in the sky, the two opposite gates in the heavens – one by Taurus and the other by Scorpius.

Just as Egyptian funerary texts describe the darkness that the Sun traverses during the hours of the night (Pyramid Texts, Coffin Texts, Amduat, etc.), so the Babylonian hero treads the path that the Sun must follow through its darkest hours.

Gilgamesh [listened to the Scorpion-man],
To the words of [the guardian of the gate (?)].
The path (of ?) Shamash [   ]
When he had achieved one league
The darkness was dense, there was no light,
It was impossible [for him to see] ahead or behind.[5]

The entrance gate to this dark path sits by Scorpius, guarded by Scorpion people, while the opposite gate by Taurus is depicted with bull-men (or attendants wearing bull-crowns) opening heavenly portals for the sun god Shamash (Figure 5).

Figure 5. Attendants with bull crowns open portals for the Sun god: the heavenly gate by Taurus (Metropolitan Museum of Art, New York).

Combining textual and pictorial evidence, we can reconstruct the topography of the heavens as promulgated by Babylonian myth: the two opposite celestial gates stand by Scorpius (scorpion people) and by Taurus (bull men/Bull of Heaven).

Where *Hamlet's Mill* Falls Short

Several of the astronomical elements relevant to this celestial topology can be found in *Hamlet's Mill*. The intersections of the Milky Way and the ecliptic are "crisis-resistant" – they are not subject to change due to Precession – and that's where souls "change trains" in the sky. [6] Macrobius is invoked, as are the gates at the heavenly intersections that he describes. [7]

What is not found in *Hamlet's Mill* is that the Babylonian myth of Gilgamesh crystallized and encoded information about the locations of the Gates to the Netherworld – one by Taurus, the other by Scorpius.

The Sumerian Gilgamesh

Several Sumerian stories about Gilgamesh survive on clay tablets, if only in fragments. One of these, 'Gilgamesh, Enkidu, and the Netherworld,' would be reworked as the incongruous Tablet XII of the Babylonian epic[8] – demonstrating the angst about the afterlife that led to the quest for immortality in the standard version.

Another Sumerian myth, 'Gilgamesh and the Bull of Heaven,' links Gilgamesh to the constellation Taurus, where one of the intersections in the heavens sits – one of the Gates to the Afterlife.

That celestial location by Taurus can be seen at the top of the plaque on the so-called King's Lyre from the Royal Tombs of Ur that now resides in the University of Pennsylvania Museum. At the top, a heroic figure wrestles with bulls – Gilgamesh and the Bull of Heaven, or the constellation Taurus. In the bottom register, we find the scorpion-man, the southerly constellation Scorpius that guards the opposite intersection in the sky (Figure 6).

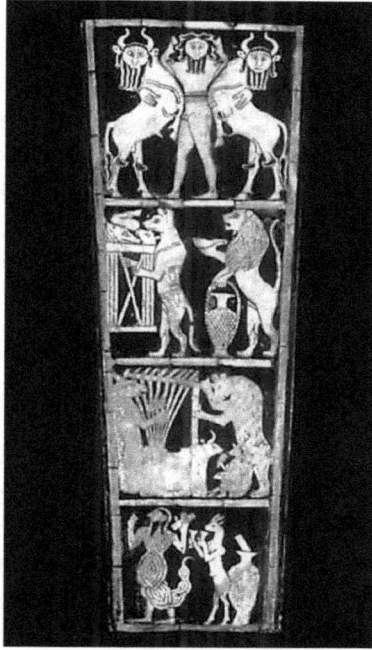

Figure 6. Plaque on the King's Lyre ftom the Royal Tombs of Ur (c. 2700 BC). At the top, a hero wrestles bulls, while a scorpion man stands at the bottom: the opposite constellations Taurus and Scorpius (University of Pennsylvania Museum, Philadelphia).

The opposite astronomical locations that would be encoded in the Babylonian story of Gilgamesh (c. 1200 BC) are already found on this Sumerian artwork from a millennium and a half earlier.

Conclusion

The myth of Gilgamesh survived for thousands of years because people of different cultures could look up at the sky and point out the story's various elements. The tale is a porte-manteau, a carry-all, that contains specific astronomical information that could be verified across differing languages and backgrounds, laying a foundation of shared understanding.

"See those twin stars?" an ancient priest might ask. "They

stand for Gilgamesh and Enkidu. The mother of Gilgamesh adopted Enkidu as her son, thereby making them brothers. The Twins (Gemini) stand facing the Bull (Taurus) – that's Gilgamesh and Enkidu fighting the Bull of Heaven at one of the intersections in the sky. They kill it and sacrifice its heart to Shamash, the Sun whose path in the sky gives the constellations of the Zodiac."

"When Enkidu dies due to the anger of Ishtar (the planet Venus), Gilgamesh travels to the gates guarded by Scorpion folk (Scorpius), to the opposite intersection in the sky. Following the night path of the Sun in the underworld, he undergoes ordeals that will make him wiser in the ways of the heavens."

For the average man, the myth of Gilgamesh was an adventure story, a rollicking action tale. But to those of keener insight and curiosity, the astronomical underpinnings could be pointed out and the theological implications revealed. Gilgamesh traveled along the path of the celestial ones on his epic journey, and there in the night sky stand the opposite Gates of Heaven that he visited – one by Taurus, the other by Scorpius.

For these mythic exploits, Gilgamesh would become judge of the dead in Babylonian lore, forever standing guard at the portals of the afterlife: "By the Ur III Period, he was regarded as king and judge of the netherworld, the role in which he was best known in Mesopotamian magic and religion in the first millennium."[10]

That's Gilgamesh at the Gates of the Netherworld.

NOTES

1 Dalley 2000: 96
2 Tigay 1982: 11
3 De Santillana, von Dechend 1969: 323
4 Stahl 1990: 133
5 Hornung 1999
6 Dalley 2000: 98
7 De Santillana, von Dechend 1969: 244
8 De Santillana, von Dechend 1969: 242
9 Tigay 1982: 26
10 Tigay 1982: 14

Bibliography

Dalley, S. 2000 [1989]. *Myths from Mesopotamia: Creation, The Flood, Gilgamesh, and Others.* New York: Oxford University Press.

De Santillana, G., von Dechend, H. 1969. *Hamlet's Mill: An Essay on Myth and the Frame of Time.* Boston: Gambit.

Hornung, E. 1999. *The Ancient Egyptian Books of the Afterlife*, (translated from the German by David Lorton). Ithaca: Cornell University Press.

Stahl, W. H. 1990. [1952] *Macrobius: Commentary on the Dream of Scipio.* New York: Columbia University Press.

Tigay, J.H. 1982. *The Evolution of the Gilgamesh Epic.* Philadelphia: University of Pennsylvania Press.

List of Illustrations

Front cover, top: Roman Republic denarius struck by Manlius, c 107 BC, (Seaby, H. A. (1989) *Roman Silver Coinage, Vol. I, Republic to Augustus*, p. 60, Manlia 1) showing the Sun god in his quadriga, the crescent Moon on the right and the heavenly X on the left, while a star at left and at right suggest the Gemini, the Dioscuri who guard the heavenly gates.

Front cover, bottom: Imprint of a cylinder seal from Babylonia, c. 1100 BC, with a presentation scene: a priestess leads a supplicant to an enthroned god who sits below a crescent Moon.

Page 3. Reverse of a coin of Delmatius, c. 336 AD, showing the combination of the pagan ladder of the planets and the Christian Chi Rho that now marks the heavenly gates (*Roman Imperial Coinage*, Vol VII, 398 for Arles).

Page 5. Fig. 1. Coin of Constantine showing standards of the Roman army, with the planetary ladder to the heavens (*RIC* 219, Siscia).

Page 6. Fig. 2. Denarius struck by the moneyer Manlius, c. 107 BC (*RSC* I, Manlia 1).

Page 6. Fig. 3. Coin of Constantine showing Sol Invictus presenting the celestial globe with intersecting lines to the emperor (*RIC* 128).

Page 7. Fig. 4. Coin of Delmatius, c. 336 AD (*RIC* Vol VII, 398 for Arles).

Page 8. Fig. 5. Coin of Constantius II, c. 340 AD (*RIC* VIII, 301 Siscia).

in the Vatican.

Page 39. South Italian kylix for use at a wine-drinking symposium. The inscribed pattern of seven leaves suggests a Pythagorean connection.

Page 41. The Harmony of the Planets notated by Johannes Kepler in *Harmonice Mundi*, as shown in *On The Shoulders of Giants* by Stephen Hawking.

Page 43. The Celestial Twins, the Dioscuri – the twin stars Castor and Pollux – on a denarius struck by the moneyer Servilius, c. 136 BC (*RSC* I, p. 88).

Page 44. Dioscuri riding right on a denarius struck by the moneyer Antestius, c. 146 BC, with a small dog (Canis Minor) running below the Gemini, Castor and Pollux (*RSC* I, p. 14).

Page 46. Coin of Maxentius (c. 309 AD) showing the Dioscuri standing guard over the twins Romulus and Remus, the founders of Rome, suckled by the Lupa Romana (*RIC* 16).

Page 47. Top. Commemorative VRBS ROMA coin, c. 337 AD, showing the Lupa Romana suckling the Roman twins Romulus and Remus while the twin stars of Gemini shine above them. Between the two stars, the Christian Chi Rho symbol (*RIC* 385).

Page 47. Bottom. Coin of Julian II, the last initiated emperor, who tried to revive the ancient rituals, c. 368 AD (*RIC* 167).

Page 50. *Consecratio* denarius of 141 AD, with Ceres welcoming Diva Faustina into the company of the celestial gods (*RIC* 382).

Page 51. Fig. 1. Left: Coin of Menaion, Sicily (2nd cent. BC). Obverse: Veiled head of Demeter. Reverse: Crossed torches (*SNG* Cop. 384)

Page 51. Fig. 1. Right: Crossed torches at Eleusis today.

Page 51. Fig. 2. Top. Coins of Metapontum with the ear of grain of Demeter. Helmeted head of Leukippos, founder of Metapontum, cross torch at left, c. 300 BC (*SNG* ANS 405).

Page 51. Fig. 2. Bottom: Ivy-crowned head of Dionysos; on reverse, cross torch at right, c. 200 BC (*SNG* ANS 590 ff).

Page 52. Fig. 3. Demeter brandishes the Eleusinian cross torch as she searches for her daughter (Apulian hydria, c. 330 BC; NY Metropolitan Museum of Art).

Page 53. Fig. 4. Top. The Eleusinian Mysteries on a denarius of the Roman Republic, c. 78 BC. Obverse: Liber (Bacchus) wreathed in ivy (Lesser Mysteries). Reverse: Ceres in chariot drawn by magic serpents, carrying torches – the Greater Mysteries (*RSC* I, Volteia 3).

Page 53, Fig 4. Bottom. Denarius of Vibius Pansa, c. 48 BC. Obverse: Bacchus (vine). Reverse: Ceres with torches, behind a plough, signifying grain (*RSC* I, Vibia 16).

Page 54. Fig. 5. Top. Roman Imperial coin with Ceres on reverse, holding symbols of the Mysteries of Eleusis. Dupondius of Claudius, c. 50 AD (*RIC* 110)).

Page 54. Fig. 5. Middle: Denarius of Domitian, c. 85 AD (*RCV* I 2636).

Page 54. Fig. 5. Bottom: Sestertius of Hadrian, c. 130 AD (*RIC* 610).

Page 55. Fig. 6. Top. Denarius with obverse of Diva Faustina, c. 141 AD. Reverse: Ceres with ears of grain and long torch, symbols of the Mysteries of Eleusis (*RSC* 134).

Page 55. Fig 6. Bottom. Reverse: *CONSECRATIO* issue, c. 141 AD, with Ceres holding torch and welcoming Diva Faustina to a blessed afterlife (*RSC* 165).

Page 57. Denarius of Antoninus Pius, c. 140 AD, showing Italia enthroned on celestial sphere with intersecting lines, indicating the heavenly gates ruled by the Roman Emperor (*RSC* 466).

Page 60. Left. Coin struck by Titus, c. 80 AD, showing the deified Octavian – Divus Augustus – and on the reverse, the imperial eagle that controls the gates of heaven located at the intersections on the celestial globe (*RIC* 459).

Page 60. Right. Denarius acclaiming Divus Marcus Aurelius, c. 180 AD, with the eagle of Jupiter perched atop a globe with intersecting lines – the gates of heaven controlled by the Roman emperor (*RIC* III Commodus 273).

Page 61. Top left. Denarius of Domitian, c. 88 AD, with his deified son sitting atop a celestial globe with intersecting lines – the gates of heaven – and playing with seven stars, the Planets (*RIC* 209a, hybrid).

Page 61. Top right. Sestertius of Antoninus Pius, c. 140 AD, with Italia enthroned atop a heavenly sphere with stars and intersecting lines – the gates of heaven controlled by the ruler of Rome (*RIC* 746a).

Page 61. Middle left. Denarius of Antoninus Pius, c. 140 AD, with Italia sitting atop a cosmic sphere with intersecting lines, the gates of heaven (*RSC* 466).

Page 61. Middle right. Denarius of Marcus Aurelius, c. 170 AD, with Providentia pointing to a sphere at her feet with intersecting lines, offering control of the heavenly gates (*RSC* 881).

Page 61. Bottom left. Coin of Constantine, c. 311 AD, with Jupiter (Iovi Conservator) presenting to the emperor a celestial sphere with intersecting lines and dotted with stars (*RIC* VI Cyzicus 80).

Page 61. Bottom right. Coin of Gratian, c. 383 AD, with Roma sitting on a throne, holding a heavenly orb with intersecting lines – the celestial portals (*RIC* 46 Antioch).

Page 64. Intaglio ring stone of a Roman initiate of a Dionysiac cult, c. 200 AD, showing a thyrsus with fillets and an X at the top and at the bottom, likely indicating the two intersections in the heavens where stand the celestial portals.

Page 66. Fig. 1. The zodiacal light rises from the horizon and envelops planets along the ecliptic – Vennus at bottom, then Mars, then Saturn at top (Photo: Tunc Tezel/TWAN).

Page 67. Fig. 2. Mosaic path at Ostia leading to a Mithraic temple, with the ladder of the planets (Mercury, Venus, Mars, etc) leading to heaven (Photo: Payam Nabarz, *The Mysteries of Mithras*).

Page 68. Fig 3. Top. Denarius of Marc Antony , c. 31 BC, with legionary standards – the eagle of Jupiter at center; left and right, standards with Moon crescent at bottom, topped by orbs or circles – the Planets in their orbits (*RSC* 30).

Page 68, Fig 3. Bottom. Coin of Gordian III, c. 240 AD, with standards showing Moon crescent at bottom, planetary orbs, and the Capricorn of Augustus sitting at top (*SNG* Cop. 526 var.).

Page 71. Fig. 4. Top. Coin of Thrace, c. 100 BC, with ivy-crowned Dionysos on obverse, and on reverse, Dionysos Soter (Schonert-Geiss 995 ff).

Page 71. Fig. 4. Middle. Coin of Metapontum, c. 250 BC, with Dionysos on obverse. Reverse: Ear of grain of Demeter with Eleusinian cross torch to right (Johnston Bronze 47).

Page 71. Fig. 4. Bottom. Denarius of the Republic, struck by V. Pansa, c. 48 BC, showing on reverse Ceres (Demeter) in biga drawn by serpents, searching for her daughter and carrying torches (*RSC* 16).

Page 75. Denarius of Marc Antony showing legionary standards with Moon crescent below and orbs above – the Planets.

Page 76. Fig. 1. Denarius of Marc Antony struck before the naval battle of Actium, c. 31 BC, to pay his legions before the engagement with Octavian's troops. On one side, the planetary standards and the *aquila* with the eagle of Jupiter, king of the gods. On other side, a Roman warship (*RSC* 30).

Page 77. Fig. 2. Top. Denarius of the Republic, c. 133 BC, with Venus riding in her heavenly biga (*RSC* Calpurnia 2)

Page 77. Fig. 2. Bottom. Denariius of the Republic, c. 131 BC, with Mars on obverse, and the war god in his celestial quadriga on reverse (*RSC* Postumia 1).

Page 78. Fig 3. Coin of Gordian III, c. 240 AD, with planetary standards topped by the Capricorn of Augustus ((*SNG* Cop. 526 var.).

Page 79. Fig. 4. Sestertius showing Divus Marcus Aurelius, c. 180 AD, with the soul of the emperor being taken to heaven by the eagle of Jupiter (RIC 659 (Commodus)).

Page 80. Fig. 5. Left. Mosaic path in Ostia leading to a mithraeium, showing the planets along the ecliptic (Photo: Payam Nabarz, The Mysteries of Mithras).

Page 80. Fig. 5. Right. Zodiacal light envelops planets along the ecliptic as it rises from the horizon (Photo: Tunc Tezel/TWAN).

Page 83. Fig. 1. Top. Drachm of Demetrios I Soter, Seleucid ruler of Syria, c. 150 BC (*SC* 1642.4).

Page 83. Fig 1. Bottom. Coin of Licinius, with genius holding bust of Sol and cornucopia, c. 312 AD (*RIC* 164).

Page 84. Fig. 2. Top. Silver coin of Valerian II, with young Jupiter atop the goat Amalthea, c. 257 AD (*RSC* 26)

Page 84. Fig. 2. Bottom. Coin of Gallienus, with goat representing Amalthea, c. 260 AD (*RIC* 207).

Page 85. Fig. 3. The constellation Auriga (charioteer) with the goat at his shoulder, Capella, sitting by the Milky Way (Planisphere: Heifetz).

Page 86. Fig. 4. Top. Denarius of Antoninus Pius with Italia atop celestial sphere with stars, c. 142 AD (*RSC* 466).

Page 84. Fig. 4. Bottom. Sestertius of Antoninus Pius with Italia enthroned upon a heavenly sphere, holding the cornucopia, c. 142 AD (*RIC* 747a).

Page. 89. Fig. 1. Kudurru of c. 1100 BC (British Museum, London).

Page 90. Fig. 2. Milky Way intersects the Zodiac between Sagittarius and Scorpius.

Page 91. Fig. 3. Opposite intersection of the Milky Way and the Zodiac, between Gemini and Taurus (Planisphere: Heifetz).

Page 92. Fig. 4. Cylinder seal impression in the New York Metropolitan Museum of Art, Neo-Babylonian period c. 700 BC.

Page 93. Fig. 5. Gate-keepers wearing bullhorn crowns open the gates of Shamash, the sun god who wears a bullhorn crown himself (New York Metropolitan Museum).

Page 95. Fig. 6. Plaque from the King's Lyre of Ur, c. 2700 BC (University of Pennsylvania Museum, Philadelphia).

www.ingramcontent.com/pod-product-compliance
Lightning Source LLC
Chambersburg PA
CBHW071745200326
41519CB00021BC/6871